Man in the Age of Technology

European Perspectives

A Series of the Columbia University Press

Man in the Age of Technology

Arnold Gehlen

Translated by Patricia Lipscomb

With a Foreword by Peter L. Berger

Columbia University Press · New York · 1980

Library of Congress Cataloging in Publication Data

Gehlen, Arnold, 1904–
Man in the age of technology.

(European perspectives)
Translation of Die Seele im technischen Zeitalter.
Bibliography: p.
Includes index.
1. Social history—Modern. 2. Technology.
3. Social institutions. 4. Social psychology.
I. Title.
HN13.G4313 301.24′3 79-23963
ISBN 0-231-04852-1

Columbia University Press
New York Guildford, Surrey

Printed in the United States of America

Contents

Foreword

THE APPEARANCE OF the first English translation of a book by Arnold Gehlen, four years after his death, is an event of some intellectual significance. Gehlen's opus is formidable. (A complete German edition is currently being published by Vittorio Klostermann, the Frankfurt publishing house, under the careful editorship of several scholars who had been associated with the author, notably Karl-Siegbert Rehberg.) From the mid-1920s to the mid-1970s Gehlen was a prolific writer, though he became well known only after World War II. His work has been highly influential within German-language social science and philosophy, though it has also been controversial. Some of the controversy has been the result of Gehlen's political position, which was considerably to the right of what is the norm among European intellectuals. Fortunately the root assumptions and insights in Gehlen's work are not linked of necessity to any political or ideological position, and indeed even in Germany Gehlen has influenced many who sharply disagreed with him politically. It is possible that the controversial character of the opus may also have something to do with style. Gehlen was an idiosyncratic thinker. He wrote with little concession to the casual reader, and he often wrote acerbically. Thus it requires some effort to gain access to Gehlen's thought. The effort is well worth while.

The centerpiece of Gehlen's work is his theory of institutions, first stated in a number of broadly philosophical writings, then applied with increasing concreteness to a large number of social problems in the contemporary world. It is quite feasible to take this theory of institutions as a sociological exercise in itself. But much of its weight will be lost if one does not perceive the philosophical roots from which it grew. These roots are in the discipline known in German as philosophical anthropology, that is, in philosophical reflection about the nature of man. In Gehlen's case this discipline was very closely linked with the findings of human biology. Quite apart from his other contributions, then, Gehlen is highly significant for the bridges he built between sociological theory and both philosophy and the biological sciences.

German philosophical anthropology has a long history, going back to Herder, Hegel, and the nineteenth-century modifications of Hegelianism (notably those of Feuerbach, Marx, and Nietzsche). Its immediate antecedents are in phenomenology, especially in the application of the phenomenological method to philosophical anthropology in the later work of Max Scheler. The biologists who influenced Gehlen most were J. von Uexküll, Adolf Portmann, and F. Buytendijk. The first systematic utilization of this biological work for philosophical anthropology was made by Helmuth Plessner, who, like Gehlen, was trained both in philosophy and biology. Plessner was also the first to explore the sociological implications of this biological research (*Die Stufen des Organismus und der Mensch*, 1928), but he did not develop a theory of institutions from it.

The biological starting point of Gehlen's theory of institutions is man's special position in the animal world. The aforementioned biologists all emphasized the unique characteristics of *Homo sapiens* as compared to his nearest mammal relatives. In this, incidentally, they as well as Gehlen differ sharply from both the ethologists (foremost among them Konrad Lorenz) and the current crop of "sociobiologists," who emphasize the commonalities between man and other mammals. Portmann's work on human fetal development is very important in this connection, suggesting that essential steps in the development of the human organism take place after the individual's birth, thus making the developing organism uniquely susceptible to the influences of the extraorganismic environment. This "unfinished" character of the human organism at birth is closely related to its unspecialized instinctual structure. Man, unlike all other mammals, is characterized by "instinctual deprivation" (*Instinktarmut*). Because his instincts do not provide him with a stable structure within which to move, man faces an "open world"; put differently, man is characterized by "world-openness" (*Weltoffenheit*). In consequence, man's condition is marked by great instability. This condition is biologically (and, by extension, psychologically) intolerable. Therefore, man himself must construct stable structures by his own activity. Social institutions are the core of this activity of "world construction."

Institutions are the culturally produced forms by which human life is given coherence and continuity, filling the "gap" left by man's "instinctual deprivation." They provide relief (*Entlastung*) of the tensions brought on by the accumulation of instinctually undirected drives. They furnish a stable *background* of human life. This means that the institutionalized sector of human life can be taken for granted, acted in spontaneously and without reflection. By the same token, the institutions, by providing this stable "background" (*Hintergrundserfüllung*), open up a "foreground" for deliberate, reflective, purposeful actions.

The basic statement of Gehlen's philosophical anthropology, in which the later theory of institutions is already adumbrated, is *Der Mensch, seine Natur und seine Stellung in der Welt* (1940, with a major revision in 1950). The most important statement of the general theory of institutions is *Urmensch und Spätkultur* (1956). This work also contains a detailed discussion of the contrast between archaic and modern societies, introducing the key category of *deinstitutionalization*. Archaic institutions are broadly encompassing and highly stable, thus approximating the functions of biological instincts; consequently, they provide very strong relief (in the aforementioned sense of *Entlastung*). Modernity "deinstitutionalizes" in that it undermines the stability of institutions, narrows their compass, and thus opens up a very wide foreground in which action is undertaken as a result of reflective rationality. Gehlen's analysis of these processes is complex and sophisticated, but the key insight is that modern societies have a built-in instability factor, with far-reaching consequences for every aspect of human life in these societies.

The present work contains Gehlen's most succinct application of his general theory of institutions to the analysis of modern industrial society. It was first published in 1949, under the title *Sozialpsychologische Probleme in der industriellen Gesellschaft*. The present translation is of the revised edition of 1957, which appeared under the new title (not a very felicitous one) *Die Seele im technischen Zeitalter*. The revised edition, in paperback, has had a very wide distribution in German; it is probably Gehlen's most influential book.

The deinstitutionalization of modernity is now related more directly to specific features of modern societies, most important among these being the ones derived from technology. It is to modern technology and the forms of consciousness it engenders (called "rationalization" by Max Weber) that one must first of all look if one wants to understand how the stable institutions of archaic man are dismantled. But the emphasis in this work, as both titles indicate, is on the psychological implications of modern institutions. The key category here is that of *subjectivization*.

Archaic institutions are highly objective; that is, they are experienced as inevitable, to-be-relied-upon facts, analogous to the facts of nature. Modern institutions, by contrast, are deficient in this objectivity. They are readily seen and indeed experienced as ad hoc constructions, here now and possibly gone tomorrow, in any case not to be taken for granted and always open to radical change. The modern organization is the typical form of this greatly attenuated institution (indeed, Gehlen doubts whether the term institution, in the sense explicated in his theory, applies here at all). Another way of describing this process, still in Gehlen's terms, is to say that modernity greatly shrinks the background of human life and correspondingly expands the foreground. Thus modernity is characterized by constant innovation, rationality, and reflectivity, and by a corresponding sense of the unreliability and changeability of all social order. This transformation, however, is not just something taking place externally, outside the consciousness of individuals. It also affects consciousness, and vitally so.

The reason for this is made clear by Gehlen's philosophical anthropology: While modernity has greatly changed man's social environment, it has not changed his nature. The same biological species of *Homo sapiens* exists and must come to terms with the new environment. That, however, is very difficult. For man continues to be instinctually deprived, in need of stability and of reliable guideposts for action. If this stability cannot be produced for him by society, he must somehow seek it within his own consciousness. Another way of putting this is as follows: Insofar as modernity reduces the objectivity of the social order, it *ipso facto* accentuates human subjectivity. This is what Gehlen calls *subjectivization,* a profound process of turning inward. This process is the anthropological and sociological foundation of a wide range of often noted modern developments—the turn to subjectivity in modern philosophy, literature, and art, and more generally the development of the ideal of the autonomous individual and his rights. It would be a misunderstanding, though, to look upon this process as occurring only in the history of ideas—such as in the historical genesis of individualism as an ideology. It is not just that modern man thinks of himself as having a very complex subjectivity; he does, empirically, have this subjectivity. Thus subjectivization is not merely a change in the realm of ideas, but an empirically available change in the realm of psychology. To that extent, at least, modern man is a *novum* in human history.

Gehlen's discussion of psychoanalysis in the present volume clarifies this matter. On the level of philosophical anthropology, Gehlen must be critical of Freud's theories. But on the level of the social psychology of modern man, there is much to be said for the same theories. They do in fact account for specific features of modern man, even if they fail to account for human nature as such. Thus one might say that the existence of the "unconscious" is doubtful if one talks about archaic men, who had a very different experience of the relation between "outside" and "inside" from modern men. But the latter may well be said to have an unconscious, and this new acquisition, if one may call it that, can be directly traced to the aforementioned transformations in the order of institutions. Archaic man had "character," a firm inner structure corresponding to the reliable structure of the social world. Modern man has a "personality," as fluid and unreliable as the deinstitutionalized social world of modernity. This personality is very complex, hard to integrate, and plausibly inclusive of the luxuriant growths uncovered by Freudian theory.

This does not mean that modern man can exist without any institutions at all (as, very logically, has been the Utopian dream of various radical neo-Freudians). That would only be possible if human nature as such had changed, and there is no evidence for this. Rather, the modern world produces quasi-institutions (also called "secondary institutions" by Gehlen), which are constantly changing, and mass-communicated schemes by which individuals can put some measure of stability into their lives. Psychoanalysis itself, and all the psychotherapies deriving from it, may be seen as belonging to this category. Such quasi-institutions do provide some support for the individual, but they are vastly more fragile than the institutions of archaic society, and this fragility is fully present to consciousness. Not surprisingly, modern individuals tend to be very nervous. Take a fundamental component of human life such as sexuality. Nonhuman mammals have no problems in this area, because their biological constitution firmly organizes their sexual drives. But archaic man had no serious problems either, because his institutions were reliable enough to allow him to organize his sexual drives with a high degree of stability. Modern man, by contrast, is beset with sexual problems, even problems of sexual identity. The confusion can be mitigated by psychotherapy, by the sexual imagery diffused through the mass media, even by cults and movements propagating this or that "sexual life-style." Typically, though, these are incomplete and transitory solutions to the problem. Thus individuals frequently change therapies, follow one image in one part of their lives and another in another part, and change "sexual life-styles" as they change other consumer preferences.

Gehlen's analysis of modern subjectivity is pushed further in other works of his. One of the most interesting is his book on modern art, *Zeit-Bilder* (1960). Further developments may be found in separate volumes of collected essays—*Anthropologische Forschung* (1961), *Studien zur Anthropologie und Soziologie* (1963)—and in *Moral und Hypermoral* (1969), one of Gehlen's less satisfactory works, much of it occupied with polemics against resurgent German neo-Marxism. There is now a growing secondary literature, mostly in German. Although Gehlen cannot be said to have founded a school, his influence is wide and growing.

The publication of the present volume in this country will for the first time allow English-speaking social scientists and others to have direct access to an exciting body of work. The publication comes at a good time. At least in the social sciences in America there now exists a much more pluralistic situation than was the case a few years ago. Orthodoxies have been weakened, new approaches have gained adherents and attention, and there is more openness to innovative departures. The appearance of Gehlen's thought in this situation has potentially far-reaching possibilities.

Gehlen's theory of institutions, with its foundations in philosophical anthropology and human biology, is likely to be one aspect of his thought that will attract attention here. For some years now there has raged a virulent controversy over the new directions taken by "sociobiologists." Despite its enraged political overtones, much of this controvery has been a tired reiteration of the old nature/nurture argument. Gehlen suggests a very different approach to these questions, rooted in human biology, yet at the same time rigorously distanced from every form of biological determinism. Further, Gehlen's theory is highly relevant for all those (a growing number) who would reestablish the old links between the social sciences and philosophy. And still further, Gehlen suggests a thoroughly original social psychology, neither Freudian nor behaviorist in inspiration. This social psychology, as Gehlen himself understood, links up very plausibly with the American tradition in social psychology deriving from William James and George Herbert Mead. A number of most interesting theoretical syntheses are conceivable in these confluences of European and American intellectual traditions.

The present volume provides access to all these aspects of Gehlen's thought. But its major thrust, of course, is the analysis of the psychological dimensions of modernity. Its original and productive method of undertaking this analysis is bound to command attention, even if some of that attention may well be critical. There has been a lot of comment on the peculiarities of modern consciousness and the modern psyche, but there has also been a remarkable lack of theoretical clarity as to what these peculiarities are supposed to be and to mean. Gehlen provides such clarity, and he does so in a manner that is perfectly compatible with the empirical investigation of these phenomena. To put it in the language of American social science, Gehlen's analysis is, at least in part, "researchable."

Gehlen offers a striking understanding of the fundamental transformation in the human condition brought on by modernity. He bridges sociological and psychological analyses of this process in an unprecedented fashion. His work thus constitutes a major step toward a general theory of modernity.

It is very clear that Gehlen's approach to modernity is critical. He analyzes modernity in a "value-free" manner, but he makes it very clear that he does not like what he sees. His stress on order and stability as critical norms has, with some justification, been associated with his overall conservative stance. It is noteworthy, however, that critics from the left have disliked the contemporary world for rather similar reasons, except that they (unlike Gehlen) identify the ills of modernity with a capitalist economic system. It would not be the first time, of course, that critics of modernity from the right and from the left have more in common than may at first be apparent. Yet it is not necessary to share Gehlen's antimodern animus in order to utilize his approach. At the very least, Gehlen gives a lucid idea of the human costs of modernization. It may fall to others to figure out whether these costs are offset by the gains, and if not, what modifications in the process of modernization are empirically conceivable.

PETER L. BERGER

Man in the Age of Technology

1 Man and Technique

The Organic and its Substitutes

SINCE THE TIMES of Nietzsche and Spengler a literature concerned with criticizing contemporary society and culture has flourished in Germany; and among its persistent motifs has been a polemic against technique.* This is a symptom that our own society has not yet finished its internal debate over the radical changes in its nature which are associated with the advance of industrialization. In Germany, public discussion often brings to the fore anxieties over the future state's probable similarity to an anthill; the manipulation of regimented brains; the individual's bewilderment, and the culture's decay. In this context, technique often appears in the role of the defendant, whereas it seems to enjoy remarkable popularity in the United States and in Soviet Russia. The Americans possess a widely diffused science fiction literature which enthusiastically projects technological Utopias, and takes pleasure in contemplating such extravagant conceptions as, for instance, a mastery over time which would allow one to wander, touristlike, into the societies of times long past.

It is not at all clear why in Germany we remain reluctant to concede the same rights of citizenship to technique as to other realms of culture, in spite of our achievements in technological innovation. The explanation may partly

* "Technique" is much less commonly used in English in this sense than is "technology"; but the meaning associated with the latter term is too narrow for the German *Technik*.

lie in the persistence of traditional views concerning the superiority of theory over practice, of pure over applied science. Or perhaps the old idea lives on, that the intellectual resources of idealistic philosophy can come to terms with all human problems; whereas, in reality, they are at a loss when confronted with technique. Indeed, on the whole, our philosophical concepts are in no way adequate to the circumstances of our own time. Our task here, however, is not the very large one of remedying this state of affairs, but a more modest one, to be pursued within the framework of philosophical anthropology. We shall look for objective viewpoints from which we may bring this astonishing domain of the human mind—technique—to bear upon our understanding of ourselves.

Technique is as old as man himself, for when we deal with fossil remains it is only when we come upon traces of the use of fabricated tools that we feel sure we are dealing with men. Indeed, the roughest wedge hewn of flint embodies the same ambiguity which today attaches to nuclear energy: it was a useful tool, and at the same time a deadly weapon. The modification to his own ends of things originally found in nature is an activity of man connected from the beginning with his struggle against his fellow man; and only quite recently have we sought to undo this fateful connection. If this endeavor is to succeed and produce perpetual peace, it must presuppose a very high level of technical attainment, without which no effective mutual control of armaments can be achieved.

Further reflection throws light upon this involvement of man with technique. Building upon Max Scheler's work,[1] modern anthropology indicates that man, lacking specialized organs and instincts, is not naturally adapted to a specific environment of his own, and is thereby thrown upon his ability to transform intelligently *any* preconstituted natural conditions. Poorly equipped as he is with sensory apparatus, naturally defenseless, naked, constitutionally embryonic through and through,* possessing only inadequate instincts, man is a being whose existence necessarily depends upon *action.*[2] On the strength of such considerations, such authors as W. Sombart, P. Alsberg, J. Ortega y Gasset, and others, have derived the necessity of technique from the limits of man's physical potential.[3]

Thus, among the oldest artifacts we find weapons, which are not given to man in the form of organs; fire should also be thought of in this connection, having come into use both for security and for warmth. From the beginning this principle of organ substitution operated along with that of organ strengthening: The stone grabbed to hit with is much more effective than the bare fist. Thus, next to *replacement techniques* that allow us to perform beyond the potentials of our organs, we find *strengthening techniques* that extend the performance of our bodily equipment—the hammer, the microscope, the telephone reinforce natural abilities. Finally, there are *facilitation techniques,*† operating to relieve the burden upon organs, to disengage them, and finally to save effort—as when use of a wheeled vehicle replaces the dragging of weights by

* In *Der Mensch* Gehlen argues, on evidence from comparative embryology, that the human gestation period is too short (by several months) to bring the human fetus to the same level of maturity at birth as for the fetus of closely related species.

† We translate the German *Entlastung* as "facilitation." *Entlastung* is a key term in Gehlen's philosophical anthropology. It characterizes the human being, as compared to other animals, as "burdened" (*belasten*) with the necessity of making arrangements for its own survival, due to the insufficiently tight fit between human physical equipment and the environment. It is thus the task of those arrangements to relieve or facilitate (*entlasten*) man's existence.

hand. If one flies in an airplane, all three principles operate—the plane supplies us with the wings we do not possess, outperforms all animal flights, and relieves us of making any contribution whatever to our own motion over vast distances.

Ultimately, all attainments of the human mind remain enigmatic; but the enigma would be all the more impenetrable if not seen in connection with man's organic and instinctual deficiencies; for his intellect relieves him from the necessity to undergo organic adaptations to which animals are subject, and conversely allows him to alter his original circumstances to suit himself. If by technique we understand the capacities and means whereby man puts nature to his own service, by identifying nature's properties and laws in order to exploit them and to control their interaction, clearly technique, in this highly general sense, is part and parcel of man's very essence. It truly mirrors man—like man himself it is clever, it represents something intrinsically improbable, it bears a complex, twisted relationship to nature.

These features are illustrated by the fact that the earliest and most fundamental technical attainments were achieved without reference to models given in nature. This is true of the starting of fires by friction of wood upon wood, of the invention of the bow and arrow, and above all of the use of the wheel, the rotating movement around an axis. This invention is so abstract that it was not attained even in high cultures, such as those of pre-Columbian South America, which possessed elaborate literatures, complex state apparatuses, and highly developed religions, and yet had to make do without the cart or the potter's wheel. Equally unprecedented in nature is propulsion by means of an explosion, as is one of the very oldest inventions—that of the flint knife, which goes back to the Günz-Mindel interglaciation, half a million years ago. G. Kraft has pointed out that nowhere in nature do we find anything like a sharp blade which, propelled in a given direction, produces a straight or a curved cut.[4]

The world of technique, then, embodies the features we associate with our images of a "great man." Like that man, it is inventive, resourceful, life-fostering and at the same time life-destroying, involved with primeval nature in a complex relationship. Technique constitutes, as does man himself, *nature artificielle.*

Over the ages, the tendency to replace missing organs has reached beyond the sphere of the body, and penetrated into deeper and deeper organic strata. The replacement of the organic by the inorganic constitutes one of the most significant outcomes of the development of culture. There are two aspects to this tendency: artificial materials replacing those organically produced; and nonorganic energy replacing organic energy. As to the former, the development of metallurgy constitutes a cultural threshold of the first magnitude; we speak of the Bronze Age, Iron Age, etc. Metals replace and outperform materials immediately available in the environment, particularly stone and wood. As late as the Middle Ages ships, bridges, vehicles, and tools were largely made of wood, and no other fuel was known. Today, concrete, metals, coke, coal, and numerous synthetic materials have largely supplanted wood, and car bodies made of plastic may soon replace those made of steel. Leather and hemp have been replaced by steel cables, wax candles by gas or electricity, indigo and purple by aniline dyes, nearly all natural drugs and medicinal herbs by synthetic products. As Freyer has argued, the ultimate goal seems to be to produce materials with certain selected properties.[5] Thus the chemist says, "I want to produce a substance which can be molded at first, but then hardens spontaneously; another which would remain plastic at no matter what temperature; a third which might be carved at will, and a fourth one which can be spun into very fine thread."

As to the other aspect of this tendency, that is, inorganic supplanting organic energy, with the steam engine and the internal-combustion engine civilization has become dependent upon underground supplies of coal and oil. Ultimately these too are legacies of past organic life, yet they entail a key transition: as far as energy sources are concerned, mankind has made itself independent of those that are renewed from year to year. As long as wood remained the most significant fuel material, and the work of domestic animals the most important source of energy, the advance of material culture, and thus ultimately population growth, met a *limit* of a nontechnical kind that rested upon the slow tempo of organic growth and reproduction. By building hydroelectric power stations and by gaining control over nuclear energy, man has freed his energy supplies from the limitations of the renewal of organic substances.

The tendency, which characterizes the progress of technique, from the substitution for organs to the replacement of the organic as a whole, is ultimately rooted in a mysterious law pertaining to the realm of the mind. Briefly put, this law is: Nonorganic nature is more knowable than organic nature. Bergson has duly emphasized this.[6] Our capacity for rational thinking, and the abstract models and mathematical concepts which it produces, approximate the givens of inorganic nature with astonishing exactitude; whereas, in spite of all progress in organic chemistry, we are not much better informed than were the earliest philosophers of classical Greece as to the real nature of life. According to Bergson, intellect can only be judged in relation to action, and its primary aim is the production of artifacts: "Therefore . . . we may expect to find that whatever is fluid in the real will escape [the intellect] in part. Our intelligence, as it leaves the hands of nature, has for its chief object the unorganized solid."[7]

The basic knowability of inorganic nature and the stubborn irrationality of the organic are remarkable facts in themselves; but it is even more remarkable that only very recently has man learned to represent the course of natural events as a dead, wholly material, yet uniform process. One can conceive of nature as "an external world of facts," as a realm of things, of properties, and of regular transformations affecting them, a realm legitimized exclusively by virtue of being there and occurring in a certain fashion.[8] This world saturated with facts, accounted for on equally factual grounds, constitutes *one single* complex, sufficient unto itself and legitimized by its sheer existence and its factual properties. Such a conception was occasionally put forward by early Greek philosophers, and made its reappearance in the seventeenth century with the rise of exact, experimental, natural science. This outlook need not be construed as an express philosophical theory (which as such would stand near positivism or materialism); rather, it generalizes an attitude inherent in scientific research and in technical practice from the time they began to be engaged in as distinctive activities. (There is a distinction between those unarticulated presuppositions which underlie the actual conduct of people, and the properly theoretical views those people may consciously expound.)

These last considerations are basic to the following

argument: the supplanting of organic by inorganic materials and energy through the development of technique is grounded on the fact that the realm of inorganic nature most easily offers itself to methodical, rational analysis, and to the associated practice of experimentation. The biological realm and that of the psyche are incomparably more irrational. Technicians and natural scientists tend to shape their own world views in accordance with the positivism-of-facts described above. The more successful sciences and techniques exercise a kind of radiation effect upon our image of the world.

In spite of this, this view of the world has been current only over the last three centuries, although mankind began to produce by means of technique some half million years ago.

The Modern Age: Its Superstructure

We all sense that since the times of stone tools or of the bow and arrow a *qualitative* transformation has occurred in what we call technique. But this change should not, as is often the case, be thought to consist in the transition from the simple tool to the machine. If we call "machine" any material arrangement which transmits energy and performs work, we can apply this term to the hunter's trap with its triggering mechanism, which has existed since the Stone Age. Even a rotating movement backward and forward can already be found in the Stone Age fiddle drill, and the continuous rotating movement of a working machine (water wheel) goes back to Roman times. Thus the difference between tool and machine is not the key qualitative difference involved in the transition from premodern to modern technique.

We get closer to the truth if we cease to visualize single machines, utensils, or discoveries, and consider instead structural changes in whole areas of culture. In the seventeenth and eighteenth centuries the natural sciences attained the modern configuration, that is, they become analytical-experimental. Briefly, an experiment consists in isolating natural processes in such a way as to make them open to observation and measurement. In two senses this makes the natural sciences, which previously relied basically on occasional observation and on speculation, similar to technical practice. In the first place the tools of physical experimentation are comparable to machines, although they are intended not to produce useful effects but to bring about pure, isolated natural phenomena. Even the inclined plane, used by Galileo to study the fall of bodies, is a "simple machine" of this nature. In the second place, by means of the logic of experimentation one isolates a natural process (which one observes under varying conditions), and to that extent the experiment constitutes a first step toward the technical use of that process. In this way, two spheres of culture, which formerly came together only in a few fields (in particular, in the construction of navigational devices, optical instruments, and precision weapons), and for the rest had remained essentially separate from one another, are now brought into the closest methodological connection. Technique derived the breathtaking tempo of its advance from the new natural science, and the science acquired from technique its practical, constructive, unspeculative bent.

Nonetheless, the astonishing accomplishments of the modern era would not have been achieved without the intervention of a third factor, the contemporary emergence of the capitalist mode of production, whose spirit, as Max

Weber showed, is a further product of the seventeenth century.[9] The invention, or rather the radical improvement, of the steam engine by James Watt was financed by a capitalist interested in its industrial exploitation. Either entrepreneurs or states interested in techniques of warfare (note, for instance, the early employment of the wireless telegraph by war fleets) made possible experimental discoveries and their practical applications.

Today it is vital to understand the functional connection between natural science, technique, and the industrial system. Scientific research employs ever-new technical devices; nature is forced open through technique. The scientist must reach an understanding with the technician, for each problem is defined by the not-yet-available equipment required to solve it. Advances in theoretical physics, for instance, depend no less upon electronic computers than upon the brains of physicists. Measurements carried out with the cyclotron, using energies of several million electron volts, enter the values under calculation and thus the related theories. On the other hand, the larger industrial complexes possess their own research establishments. Natural science is no longer the monopoly of universities—indeed sometimes only grants from industry allow otherwise underfinanced laboratories in technological universities to keep going. The notion that technique constitutes "applied science" is obsolete and old-fashioned; today the three establishments—industry, technique, and natural science—presuppose one another. What is the ultimate basis of pharmaceutical chemistry—biochemical research, the industrial firms that commission it, or the production and marketing organizations of those firms? It no longer even makes sense to pose the question in this fashion.

Supernatural Technique: Magic

The rapid advance of modern technique has thus taken place in close alliance with natural science and the capitalist mode of production, which extend their hold with equal rapidity. All these factors feed upon one another. One cannot expect such historically unique and radical processes to remain without influence upon the consciousness of the men involved in them. The pragmatic-positivistic attitude which characterizes this "industrial system" has decisively extended its reach beyond the confines of the system within which it originally developed. It has affected, for instance, the political realm and, even more so, the realm of interpersonal relations. We shall have to deal with these phenomena later on, since they constitute the social-psychological problems of industrial society. For the time being, however, we must develop another idea, which should help to illuminate the human condition. This concerns the human impulses operating in the technical realm.

During by far the greatest part of its history, as we have seen, mankind has made do with fairly modest technical resources, however ingenious those early discoveries might have been. Such basically simple instruments and artifacts as the war chariot, firearms, or the plow, could have extraordinarily significant historical and social consequences. Even so, technique did not come to occupy the very center of man's vision of the world, and thus also of his conception of himself. This *is* what is happening today, when for instance we look at cybernetics, to the theory of techniques of regulation, for clues to the workings of our own brains and nervous systems.[10]

If we ask, "Why did this not happen before?" we get a surprising answer. For millennia, in all primitive cultures as well as in the higher ones (the Egyptian, the classical, etc.), man believed in the possibility of a "supernatural technique"—of what today we call *magic*. Since prehistoric times magic has held a central place in man's conception of the world and of himself. Even in monotheistic cultures which denied the possibility of magic, magic maintained a foothold on the margins of society—as shown by the trials of witches and magicians in the Middle Ages—and only modern, technical-scientific culture has dealt it a mortal blow.

Maurice Pradines calls magic "an attempt to bring about changes to the advantage of men, by diverting things from their own path and toward our own service."[11] It is easy to see that this definition can encompass both magic and technique proper, thus both supernatural and natural technique.

We cannot undertake, here, a closer analysis of magic,[12] but we must emphasize its tremendous diffusion in time and space. If we consider the remarkable similarities found in the magical practices of all races and civilizations, we see that magic must involve something anthropologically fundamental. "Rainmaking," for instance, was practiced in classical antiquity; according to Diogenes Laertius, Empedocles possessed this skill. The *Hammer of Witches* (1487) gives explicit directions for countermagic to be used against magically induced bad weather. New Guinea natives practice rainmaking, just as do Omaha Indians, the Bantus of Delagoa, and the Chinese.

If we consider more closely the numerous accounts and documents available, a central concern of the magical "arts" becomes evident: the need to ensure the "regularity of the process of nature," and to "stabilize" the world's rhythm by smoothing out irregularities and exceptional occurrences. Thus, when defective births, moon or sun eclipses, or other strange events appear as unfavorable "signs" against which magic must intervene, what is being sought is the reinstatement of the usual uniformities of nature, just as when magic is employed to call forth the usual rains or winds which have failed to appear. The same holds true for the innumerable examples of "fertility magic," used to ensure the cycles of vegetable life or to increase the number of plants or animals. In fertility magic it is important to respect precisely given dates, seasons, or hours, or perhaps recurring phases such as the beginning of cultivation, of sowing, or of harvesting.

This primary human interest in the regularity of the processes of nature deserves emphasis: It betrays a semi-instinctual *need for stability in the environment*. Since reality is unavoidably subject to time and to change, the most stability one can hope for consists in the same effects repeating themselves automatically and periodically, as indeed they tend to do in nature. The primeval, "a priori" conception of the world, not yet influenced by science, views the world, and the men who are part of it, as caught in a rhythmic, self-sustaining, circular process of motion, thus constituting an animated *automatism*. Also, the magical forces with which the world is filled are neither arbitrary nor spontaneous; one can set them into motion by means of the appropriate, precisely repeated formulas, after which they operate under their own impulse, necessarily and automatically.

A considerable residue of this primeval, innate view is still present in astrology, in spite of all the "rationalizing" effects of the new, scientific world image. Most of us would be astonished at the number of businessmen and politicians who believe in an inescapable connection between the immense, rotating mechanism of the stars and the destinies of individuals—a connection which the metaphysics of primitive peoples views, without any sense of contradiction, as both willed by spirit and necessary. Something that has resisted all the thrusts of offended reason must obviously be deeply rooted in the mind of mankind.

The fascination with automatisms is a prerational, transpractical impulse, which previously, for millennia, found expression in magic—the technique of things and processes beyond our senses—and has more recently found its full realization in clocks, engines, and all manner of rotating mechanisms. Whoever considers from a psychological viewpoint the magic which cars exercise upon today's young, cannot doubt that the interests appealed to lie deeper than those of a rational and practical nature. If this seems improbable, one should consider the fact that a machine's automatism exercises a fascination entirely independent of its practical uses, a fascination that might well be best embodied in a perpetual-motion machine whose only goal and activity would consist in forever reproducing the same circular motion. None of the innumerable individuals who over the centuries have grappled with the insoluble problem of perpetual motion, did so in view of any practical effect. Instead, they were all fascinated by the singular appeal of a machine that runs itself, a clock that winds itself. Such an appeal is not merely intellectual in nature, but has deeper sources.

That appeal involves what we may call a *resonance phenomenon*. Beset by the enigma of his own existence and his own nature, man must define himself by referring to what is other than himself, other than human. His awareness of himself is *indirect*, and his search for a self-definition always must consist in comparing himself to something nonhuman, and then differentiating himself from that.[13] It is not difficult to establish this point with reference to the concepts of divinity of the higher monotheistic or polytheistic religions, or alternatively with reference to the much more ancient and more widely diffused myths concerning man's descent from animal demons. Also, in interpreting his own psyche, man has largely referred to phenomena of the external world; availed himself of shadows, of blood, of mirror images and other visual phenomena in order to penetrate his own inner nature. Primitive religions have found throughout nature silent answers to the question of man's own essence.

Within this orientation, however, what necessarily makes the greatest impression is the fact that natural processes advance rhythmically and periodically, with an imperturbability that bespeaks a "logic," whether the attention be fastened upon the puzzling exactitude of the recurrent motions of the stars, or upon the stubborn, stereotyped, immutable habits of animals. And in fact, in a number of quite central aspects of his own nature man himself *is* an automatism; he *is* heartbeat and breath, he lives in and by a number of meaningful, functioning, rhythmical automatisms—think of the motions of walking, think above all of the ways in which the hand operates. Think of the "circle of action" which goes through object, eye, and hand, and which in returning to the object concludes itself and begins anew. The fascination exercised by the analogous processes of the external world bespeaks a "resonance," which conveys to man an intimate feeling for his very nature, by focusing on what echoes his nature in the external world. And if we today still speak of the "course" of the stars and of the "running" of machines, the similarities thus evoked are not in the least superficial; they convey to men certain distinctive conceptions of their own essential traits based on "resonance." Through these similarities man interprets the world after his own image and, vice-versa, himself after his image of the world.

Objectification and Facilitation

We come thus to a point of great significance for determining the relationship between man and technique. For if there is a deep-seated bond between man and those processes of the external world that advance rhythmically, periodically, under their own momentum, this makes more comprehensible the *drive components* (*Triebkomponente*) implicit in technique. There is a widespread prejudice, largely of academic origin, to the effect that technical behavior is "merely rational" and "exclusively goal-oriented." Yet, as Hermann Schmidt has emphasized, the *objectification of labor* involved in technical phenomena is the result of a process specific to mankind, but of which we as individuals are not conscious, and whose motivation flows from the "sensual side of our nature." "Any group of men placed under identical conditions would always undertake to objectify labor as in response to a drive." In this connection, Schmidt quotes a remarkable statement of Walter Rathenau's: "Mechanization is not the result of free, conscious deliberation, expressing mankind's ethical will; rather, it grew without being intended, or indeed even noticed. In spite of its rational and casuistic structure, it is a dumb process of nature, not one originating from choice."[14]

The process in question can be variously construed.

Man—as I have shown at length elsewhere—is a being constituted for *action,* for the modification of the facts of the external world.[15] One of his essential characteristics is the *circle of action* (*Handlungskreis*)—a modifiable, directed motion capable of correction on the basis of its outcome, and which in the end may become automatized and wholly habitual.[16] "Each of our meaningful operations," writes H. Schmidt, "necessarily takes this form of a self-contained circle of action, where feedback connects the subject with himself on the basis of the previous results of his action." *Each* is appropriately said, for even the speaking-hearing circle constitutes such a circle of action—and language is the vehicle of all mental activity. "The circle of action is the universal form of man's meaningful expression."[17] In keeping with this, Norbert Wiener calls feedback a very general feature of forms of behavior: "In its simplest form the feedback principle means that behavior is scanned for its results, and that the success or failure of this result modifies future behavior."[18]

It is not easy to understand the irrational impulses at work within technique. The need of man to read himself into nature and then to interpret himself in terms of nature (a need evidenced all over the world, and preserved at the very core of religion) is fundamental. All periodical, cyclical processes evoke a near-instinctual resonance in man; from the beginning he has seen himself as caught in a cycle of rebirth. Having thus brought himself close to the world, he relates to it mainly through his own power to act. Magic as supernatural technique brings into the circle of action the totality of the external world, makes it possible to summon the wind, to call forth the seasons, to transfer one's illnesses to animals. The basic need behind the practice of magic—the need to stabilize the course of the world and free it from disturbances—is the need of an *acting* being.

It is an equally primeval fact, however, that man also objectifies his own material action, and through it makes an impact upon the world; he sees his action as part of the world, allows the latter to extend and reinforce his own action; he "objectifies" his own labor. Hence the tool. The stone is a "representation" of the fist, stands in its stead, and indeed magnifies its effect. Thus the narrow sphere of one's actual control merges into the wider sphere of what one can control through imagination. In fact the expenditure of one's physical energy diminishes in relation to the masses set in motion. Working with tools is demanding, but magical formulas suffice to stabilize the weather or to guarantee the spring's return.

Here one can see in operation a further, fundamental human law: the *tendency toward facilitation*. As we have made clear elsewhere, the principle involved is one of general anthropological significance.[19] Here only its implications for technique are relevant: the "larger circle of action" of magic relieves the burden of the weakness and helplessness one feels when confronted with the powers of nature, by facilitating the reduction of the world to human dimensions. The smaller circle, that involving work, facilitates in the literal, physical sense. The "objectification of human labor" into the tool makes it evident that a lesser effort can achieve greater results; for this reason we have already discussed the use of tools as a matter of organ facilitation.

One should not forget a third process of facilitation; both techniques share the same, implicit purpose, or at any rate tendency, to build habits, to lay down routines, to make many actions a matter of course. This third tendency toward facilitation is expressed by R. Wagner as follows: "In this fashion the supreme tribunal, the cerebral cortex, frees itself time and again of whatever task has become highly probable, everyday and trivial, and keeps itself available for unusual and more sensational performances."[20]

One may now understand why technique, from its beginnings, operates from motives that possess the force of unconscious, vital drives. The constitutional human features of the circle of action and of facilitation are the ultimate determinants of all technical development. This is not to say that one may predict the content of a given invention by reference to those determinants; clearly the operation of an engine is to be understood on the basis of physical and technical considerations, rather than by reference to the motives leading up to its construction. However, if we consider the development of technique in its totality, we come upon a law more fundamental than these physical considerations, a law obeyed unconsciously yet invariably, which can only be identified on the basis of the concepts of the progressive objectification of human labor and performance, and of increasing facilitation.

This process develops in three stages. In the first, that of the *tool,* the physical energy necessary for labor and the required intellectual input still depend on the subject. In the second, that of the *machine,* physical energy becomes objectified by means of technique. Finally, in the third stage, that of *automata,* technical means make dispensable also the intellectual input of the subject. With each of these steps, the objectification of goal attainment by technical means advances, until the goal we have set ourselves is accomplished, in the case of automata, without our physical or intellectual participation. In automation, technique attains its methodical perfection, and this conclusion of a development in the technical objectification of labor which had started in pre-history, is a distinctive feature of our own time.[21]

In the course of this development, which accompanies and largely determines the history of mankind, it is only recently that technique has come to occupy the space held for hundreds of thousands of years—during the times when men knew only primitive tools—by magic, the "supernatural technique." But magic was also intended to (in the words of Pradines) "divert things from their own paths and toward our own service"; it sought unconsciously to strengthen the effectiveness, to multiply the reach of human action; and it envisaged something like the "great automatism," whose operations are regulated by the information feeding back from areas of possible disturbance.

Automation

H. Schmidt's law of the three stages suggests that, from man's standpoint, the objectification of actions and faculties into the external world develops as if from the outside toward the inside. At first it is the performance of organs which are strengthened, improved, facilitated. Then the same thing happens to physical energy inputs: the energy expenditure originally carried out organically (by animals or by man) is taken over by nonliving matter. In the third stage, where we find ourselves, what becomes objectified is the circle of action itself, including its control and direction. At the same time, that part of physiological life which operates through circular sensory-motor processes becomes objectified—as does that part in which regulation is performed in a wholly automatic fashion, for instance by means of chemically transmitted information. Finally, computing automata can solve differential and integral equations faster and more effectively than man, and appear as "a new source of mathemetical knowledge."[22]

These modern regulatory devices endowed with feedback all rest on the principle that, unlike the automobile, the system does not vary in its operations according to commands imparted from outside, but rather under the influence of the results of those operations themselves. To this end one must build into such automata sensory devices, such as a thermosensor in the hot water tank, which switches an electric current on or off according to the temperature. Here thermal quantities are being regulated, but a variety of mechanical and electrical ones can similarly be regulated. "The essential is that such a mechanism should continuously react upon itself via a closed circle. Or, as one might also say: These devices are so arranged that a very small portion of the energy stream traversing the system is put to use for the regulation of the energy stream itself."[23]

The circle of regulation * can be considered in the first place as a "copy" of the circle of action; and in fact it is technically possible to build a car where the burden of driving is taken off the driver and taken over by automatic controls. But, apart from the circle of action, the same structural principle is found to operate in many physiological regulatory processes. The regulation of blood pressure, for instance, takes place through a self-enclosed circle of operations endowed with feedback.[24] Within the walls of the larger blood vessels, the aorta for instance, there are sensitive nerves which report information about rising blood pressure to a vascular nerve center in the medulla

* *Regelkreis:* the expression adopted in the text translates literally a German term whose meaning is not very different from "feedback."

oblongata, and there activate a countereffect. The tension in the walls of the peripheral vessels is reduced; they become dilated and admit a larger flow from the aorta, where, as a consequence, the pressure diminishes. But this activates the opposite process, so that the blood pressure oscillates, pendulumlike, around a central value. Numerous biological states such as the regularity of breathing, the saline concentration and sugar content of blood, and bodily temperature, are regulated in this fashion, as is, for instance, the vestibular organ that controls equilibrium. Physiologists employ concepts also used in the realm of feedback automata, such as that of "reafference," in order to describe more effectively both voluntary and involuntary movements.[25]

A philosophical evaluation of such matters would be premature, and it is best to avoid hasty mechanistic interpretations, such as that the insights into "life" afforded us by the technical circle of regulation have now made plain the mechanical nature of life itself. All one can say is that the circle of regulation, viewed as a complex of operations, appears to share the same *form* as the human circle of action and a number of physiological mechanisms of regulation; but this allows for fundamental differences in the components of that form. Thus what we have is an "isomorphism," a similarity of configuration,[26] not a similarity of nature; and we are today no closer than before to a "synthesis" of life. This leaves open the possibility of taking certain life processes, including some of the greatest significance, and treating them as objects in the external world, inanimate, and as it were "estranged." There are other, equally significant, processes for which this is not as yet possible; though cell division has already been analyzed with reference to regulation processes.

Thus the advance of technique allows man to transfer into inanimate nature a principle of organization which operates at various points within the organism. We have spoken of inanimate nature, meaning by this not unprocessed, raw, lifeless nature, but rather technical equipment produced by man himself. Prehistoric man attributed to raw nature a principle of organization, though in fantastic form, when he employed magical techniques to address the clouds or the winds as if they could hear him.

Modern technologists, however, have developed their regulatory devices without being aware of their isomorphism with biological processes, which became apparent only later; they have somehow, *unconsciously* and semi-instinctively, produced models applicable to certain life processes. As a predictable result of this, rich areas of experience such as technology, physiology, biology, and psychology, will enter into closer and more frequent contact, exchanging queries and theories with one another. It is still too early to consider cybernetics as a distinctive, self-standing, general science. For the time being it constitutes an endeavor to consider jointly and to cross-fertilize several sciences. Sociology will have to be added to those disciplines already mentioned, since the notion of "signaling back" raises the problem of communication, or rather of information transmission, not only in machines (such as computers) but also in living beings.

2 Novel Cultural Phenomena

Abstractness

BEFORE WE DEAL with the contemporary social and cultural situation, we must clarify our views concerning technique, since to understand mankind's current condition we must take into account the alliance between technique and the industrial form of production. The "culture of machines" developed in some of the same areas of the globe as did the analytical natural sciences, the first power and working machines, and the spirit of rational capitalism—the spirit that impelled these previous developments. Yet the rapid spread of machine culture over the rest of the globe would be impossible to comprehend only on rational grounds. This is why we have insisted that the roots of technique are deep and that unconscious drives operate behind technological development. Man *must* strive to augment his power over nature, for this is the law of his existence; and, when necessary—and it has been so for millennia—he makes do with an imaginary power, such as that of magic, where real power is not attainable.

On the other hand, it is also not sufficient to account for technique by referring to man's innate striving for power (as one commonly and for that matter rightly does). With the same blind energy that propels his spirit, man also seeks to objectify himself: he finds in the external world models and images of his own puzzling essence, and uses the same faculty of "self-estrangement" to transfer his own action to the external world, allowing the latter to take it over and carry it further. Hence man's remarkable fascination with the automatism and the orderliness of circular motion, which he first observed in the skies, and with the monotony of eternal recurrence. These phenomena evoke a resonance in man's own pulse, and conversely his action is perceived as driven by the same forces as terrestrial rotation and the rhythms of nature. The world's equilibrium is seen to lie on the same plane as the circle of human action.*

If one sees how much the laws of "supernatural technique" dominated primitive thinking, one cannot expect man's spiritual existence to have been left unaffected by the transition to industrial culture. The latter transition is comparable in magnitude to the Neolithic revolution—that prehistoric time when man abandoned hunting, and opted for a settled existence based on agriculture and the domestication of animals. Later this was to lead to larger and denser settlements, differential distribution of wealth and power, the division of labor, and last but not least the emergence of accepted deities with their temples and cults. An equally profound transformation of the world will be wrought by industrial culture, when man has finally spun his wireless net over the entire planet. We are but at the beginning of this process, which has gone on for only two centuries.

Still, it is now possible to detect in the contemporary situation some features of the culture toward which we are moving. One is struck at first by the thorough *intellectualization* of the cultural fields of the arts and sciences, with a consequent loss of intuitiveness, immediacy, and unproblematical accessibility. The frontiers of the arts and

* *Die Stabilisierungsfläche der Welt verläuft in der Ebene des menschlichen Handlungskreises.*

the sciences are ever more abstract and disembodied. The larger public has become aware of this since the publication of the general theory of relativity.[1] Then, around 1916, a certain body of literature sought to popularize some mathematical conceptions of which there was no intuitive model. There was a certain reluctance to surrender the notion that the world in which people live should also lend itself to their grasp; and yet since that time not only the physicists but also the political scientists and the technologists have had to give up that assumption. To return to physics, though, we have since been taught that we have to do with a granular structure of space, with an absolute minimum time interval, and with elementary particles that change their identity. In short, these are conceptions accessible only to specialists, impossible to explain to the layman; and we must leave them to the experts.

To give another example: in fields of psychology that value rigor in concept formation, there is an increasing tendency to employ mathematical methods. Such questions as whether there is a correlation between mental disorders and physical characteristics; whether an individual's I.Q. is related to his other traits; or even whether intelligence itself is a pluridimensional construct resulting from mutually independent factors—all these questions lend themselves to analysis only on the basis of quantitative and statistical treatment. Factor analysis, a procedure for detecting the controlling factors underlying observed correlations, becomes then an "adventure of the spirit,"[2] but one which is open only to those having above-average mathematical competence and specific aptitudes. Previously, psychology made much larger use of intuitive methods and insights, yet today even psychoanalysis, which used to prefer intuitive to quantitative methods and originally argued in terms that were remarkably and shockingly accessible, is clearly becoming more and more abstract. An exemplary and (in its own terms) remarkable book in the field employs such concepts as learning quantum, cognitive restructuring, tolerance of frustration, motivational grading, and so forth; it operates, in other words, in the context of a kind of higher mathematics.[3]

This tendency toward abstract conceptualization and this preference for mathematical methods are also apparent in economics, which previously contented itself with simple and comprehensible mathematical models such as the notion of equilibrium. One encounters the same phenomenon in sociology, especially in the field of "sociometry,"[4] and even in anthropology, which previously was concerned only with description.[5] And where a discipline cannot easily move in the same direction, as in the case of history, its persistent intuitiveness makes it apt to seem unrealistic and superficial. At the very least we require today's historian to show his awareness of the multidimensionality of his subject. He must consider the sociological, economic, and psychological factors, together with the political, and this leads the historian to interpret his object within shifting frames of reference.[6] Today, attempts to account for the richness of historical events unidimensionally or even dialectically, in the manner of Hegel or of Spengler, appear dated, and give the impression of being too immediate and poetic.

Not, however, poetic in the sense of modern poetry, which has long since become intellectualized and abstract, and rejects—as Gottfried Benn phrases it—"a poem where the object of poesy and the poetizing ego are separate and counter-posed."[7] This quotation strongly reminds one of certain intriguing theories in physics, or of multivalued logic, according to which reference to the subject can itself enter into the content of a sentence. The physicist finds the naive objectification of perceptions just as questionable as the poet finds the naive objectification of states of feeling. One takes sides with Mallarmé, who says that a poem arises not from feelings but from words. This means that one maintains a distance from immediate nature, both internal and external, since, as Benn says, "in nature we find colors and sounds, but no words." This important poet has also clearly described what follows from this: once a few words or lines have suggested themselves, "one places them into a kind of observation device, a microscope; one tests them, colors them, probes their pathological spots"—a skeptical and sophisticated process, from which there issue ciphered stimuli, a "wave packet," as a physicist might say.

The stylistic similarity between this kind of lyric poetry and abstract painting or atonal music is striking, as is the air of similarity between the arts and the sciences of today. We hardly need to be told that the concave vaults of modern architecture, with their curved and double-curved surfaces, are closely connected with the geometries of Riemann or Lobachevsky. The "curved space" of physicists has made its impact: there are rotating houses where the points of reference for orientation themselves change, as prescribed in Einstein's theory.

We are not arguing for anything like a direct, or perhaps a causal, relationship between all these phenomena and the "spirit of technique." All the same one should not fail to note that such cultural changes were made possible by technique, since technique placed the whole society on foundations of concrete and steel, hid nature from sight, and made survival depend upon the most daring and improbable intellectual projects. The readiness to reverse all assumptions, the irresistible appeal of "pure" solutions, the emancipation from the obvious and those things which habit had rendered "natural"—these are the motifs at work in the modern arts and sciences. Previously, obscure and weird impulses could only operate underground. Today they are expressed loudly and openly, unrestrainedly, and pervasively, since technique has given them the external support they needed to become vitally significant components of a triumphantly successful, tangibly real, more-than-spiritual force. The antinatural effects evoked by non-Euclidean spaces or by three-valued logic, as well as by Picasso's paintings, are intended effects. Just as in the solution of a technical problem, they are the products of an unconscious logic; sophisticated, aware consciousness fancies itself to be in charge of the process, but actually the process makes use of that consciousness and irresistibly urges it forward. What is involved is one of those rare and great transformations of the human condition, which affect not just the conduct of existence or the economic structure, but the very structure of consciousness itself, the ultimate dynamic of human impulses.[8] We are today watching the operation of the human intellect emancipated from those moral impulses which the Enlightenment believed to be inherent in it, but which are now reduced to the unhappy role of seeking in vain to hold back the advance of the efficient, the functional, the technically possible.

The Diffusion of the Experimental Mode of Thought

Given the circumstances we have described, apart from the small circle of top professionals and experts, the modern arts and sciences are only accessible to a small number of interested laymen properly trained in a given field. An interest in the field, however great, is not sufficient; intense study is also required. To become knowledgeable about atonal music, behavioral research, the study of personality with its batteries of tests, or whatever clever discovery one may deal with, requires intensive and systematic application, and in most cases above-average talent as well, and is thus possible only for a few nonspecialists. There are not many professionally active and educated men, anxious to share in the events of their own time, who can afford this additional commitment of their time and energy. Many such people feel barred from pursuing such an interest, and fall prey to the untiring, systematic, and effective promotional activity being carried out on behalf of the abstract arts in particular—an activity which gains in intensity and impatience from the feeling that it cannot be successful everywhere.

As the arts and the sciences become thus esoteric, they can no longer function as substitutes for religion—a fact that has not so far gained the notice it deserves. As late as the nineteenth century, certain theories or certain artistic movements were accessible enough to the common understanding to generate mass movements characterized by distinctive world views. At one time, to be a Darwinist or a Wagnerian meant to have made a serious emotional commitment to certain theories which supplied a certain general orientation toward existence. Whether rightly or wrongly, this was at any rate possible, and in fact each major author of the time, from Schopenhauer and Nietzsche to Ibsen, Strindberg, Hauptmann, and George, intended to start a "movement" and to make a significant difference to the spirit of his times. Today no such possibility exists, owing to the disappearance of stable themes around which mass opinion might aggregate itself. Many of the individuals who crowd Picasso's exhibitions undoubtedly feel passionately attracted to this art form, have enthusiastically responded to its appeal, and will continue to pursue it. But the Church has no reason to worry over this: it can well afford to have Le Corbusier build a chapel at Ronchamp,[9] whereas it would never have allowed a place of worship to be frescoed with scenes from *Parsifal*. The Christian denominations fought bitterly against Darwinism, but they see in contemporary genetics no reason for concern, since it is of such complexity that it has long since become impossible for a single mind to encompass it. Indeed, the increasing abstractness of the arts and sciences has the effect of strengthening religion as a world view. Religion can increasingly afford to consider the new arts as being both "objectless" and unobjectionable; and the churches can avail themselves of them with the same ease with which they use television, radio, or the telephone. Philosophy, too, has ceased to compete with religion in the realm of world views, since politics and natural science have become autonomous, and can no longer be gathered under a single philosophical umbrella. This phenomenon reveals what deep changes have occurred since the Enlightenment—what would Kant have been without the French Revolution and without Newton?

Thus, the inner bond between modern intellectual culture and technique estranges them from religion, thereby leaving to the latter the sphere of world views proper. But here we want to consider the relationship between the former two, and to establish one particular meaning of the word "technique." The central problem in both fields is that of feasibility,* of the *uses* to which certain methods can be put.[10] This entails a complete reorientation of inquiry. It is no longer a matter of starting from predefined ends and from given research themes, and seeking the best technical means to the former or the research methods most appropriate to the latter. Quite the reverse: it is a matter of varying the ways in which we represent things, our cognitive devices and search procedures, in order to test them thoroughly, to discover what they are capable of, and to see to what results they lead. Of course, with respect to technique we still start with given ends in view, and seek the best means for realizing them; for instance, one seeks to curb the noise of an engine, to generate in a firm the "working climate" that will yield the maximum gratification and work satisfaction (in the psychological sphere, this is also a technical problem).

But there is an increasingly significant approach to the formulation of problems where the question *is* what unforeseen outcomes will result from a given process. In this way, uses of electricity, of the electron microscope, and of atomic energy have been explored in the most diverse fields. In this context the word "technique" preserves something of its original meaning: the art of contriving, canniness, the success which flows somewhat unexpectedly from effort, and is then mastered. It is ultimately a matter of what one can accomplish by means of given techniques, of methods (including those of an intellectual nature) that one varies without a preestablished end in view, but through restless experimentation. Thus for instance the

**Machbarkeit*: an expression intended to indicate the open-endedness of arrangements as well as of things—the assumption that they can all be "made and unmade" at will, since they possess no intrinsic justification.

discovery of X-rays was put to use not just in the production of radioscopic images, but also in the treatment of living tissues, and in the analysis of works of art.

Understood in this sense, the technical element acquires great importance in the arts and sciences, where one makes endless recourse to experimentation and to methodological devices. This remarkable fact is best illustrated by the development of modern painting. In the decades that see its development out of impressionism, by way of Cézanne on the one hand and Seurat on the other, into expressionism, "abstract" painting, and the surrealism of Max Ernst and Dali, there operates an unmistakable logic, which first appears as the dissolution of the object. But this in turn is only the result of deeper transformations. On the one hand certain contemporary psychological theories made a deep impression upon painters, changing their conception of the basic features of optical-pictorial processes. This is already evident in the way impressionism, following the then-dominant psychology of sensation and of association, dissolves tangible reality in color. The influence of scientific psychology, already unmistakable in Seurat's "pointillisme," is even greater in expressionism and surrealism, which are quite unthinkable apart from the impact of psychoanalysis. An alternative road led to the determination of distinctive "image components" (*Bildelemente*), whether conceived as configurations or as color; and here contemporary abstract painting shows the ultimate point of arrival.

Even this short exposition suggests that these variously associated views and theories were treated essentially as methods, to be pursued until they had yielded all their possibilities. But the possibilities proliferated as if by a logic of their own, leaving the artist not in control of them, but trying to keep up with them. The whole history of modern painting suggests the imagery of a permutation of possibilities, as when chemists conduct exhaustive series of experiments. Once a new combination has yielded its distinctive effect, it is laid aside and never taken up again. This endless experimentation is most apparent in Picasso, whose countless paintings suggest not the search for paths to a given target, but rather a relentless experimenting with methods in order to discover what they lead up to. All this expresses a scientific as well as a technical inspiration; and all good modern artists (those, that is, who know what they are doing) continually verify Benn's metaphor about placing words under the microscope, testing them, coloring them, and finally putting forward what results.

As we have suggested previously, the experiment is the connecting link between technique and natural science. Wherever one inquires into the constant and the variable aspects of an object, consciously varying the latter; wherever one replaces previously accepted axioms with different ones, or postulates counterintuitive principles, in order to explore their implications; wherever one detaches a method from its original context and explores its outcomes within a different one—in all these situations the same experimental inspiration is at work. We are convinced that it was first developed within technique and the sciences; but later, thanks to their enormous success, it became deeply rooted in social reality; it became, as it were, a power unto itself, until today it has become the progressive form of consciousness of the age, and asserts itself everywhere with the irresistibility of fate. But in this context what matters is the *form* of this consciousness, not its "mechanistic" content. It is immature and provincial to argue with the "spirit of the natural sciences" or of technique on the basis of mechanistic models of thought. The modern spirit is in principle indifferent to content, being primarily concerned with the question of *how* a thing can be brought into existence. Thus, psychology, sociology, and economics are abandoning their descriptive-objective phase, leaving behind "mechanistic" as well as "organic" or "olistic" models, as constituting problems meaningful only within the framework these disciplines have relinquished. What they want instead is to become mathematical, and thus to avail themselves of the enormous wealth of formalized tools for thought—the contents will then take care of themselves!

We find the same spirit also at work within architecture, with the reversal of the traditional axioms that required a house to stand on foundations and to have four corners, and a vault to be convex. We now find round or oval houses, houses on stilts, gondola-rooms hanging in the air, undulating vaults. Thus far the process seems outstandingly simple, but the resulting problems and innovations are fascinating, in view of the static calculations involved, and of the possibility of exploiting hitherto unexplored properties of materials. "Beginning from this point, principles previously applicable only to extremely light materials—membranes which hang like a canopy, nets woven like hammocks, surfaces as tensed as drums, the principle of the soap bubble, whose molecular tensions hold themselves in equilibrium—can be translated into carrying structures which support and enclose at the same time."[11] This quotation also makes clear how little one understands the new arts if one fails to go beyond their appearances and to confront the experimental problems they grapple with.

When this experimental attitude pervades the cultural sciences, utterly new phenomena make their appearance, and cause confusion even among specialists. For clearly the once-classic methodological distinction between the natural and the cultural sciences is disappearing—although it is also true that some disciplines, such as sociology and economics, could never easily accommodate that distinction on account of their objects. One of the most significant pioneers of this procedure, Vilfredo Pareto, expressly introduced the "logico-experimental method" into sociology, thereby producing a phenomenon previously known only in the case of physics; namely, it became impossible to identify the object of discourse apart from the method of discourse. Pareto evaluated all manner of views (philosophies, dogmas, statements of world views, ideologies) but not according to their truth-value, and indeed denied once and for all their claim to any such value. Rather, he interpreted them as virtual actions, that is, as incipient acts. Through statistical analysis of a large number of such acts, he distinguished the variable from the constant elements, and viewed the latter as basic impulses to action, that is as instincts. Pareto considered the intellectual historians' traditional treatment of these ideologies, dogmas, etc., as itself being a special case of ideological behavior, and treated it as any other case—in the same way in which recent physical theories treat classical mechanics as a special case.[12]

We cannot go beyond this brief presentation of Pareto's method to the incredible wealth of insights which it unexpectedly generated. At any rate, from the outside, with the intellectual resources of everyday understanding, it simply was no longer possible even to indicate what was the object of discourse, as when, for instance, Pareto analyzed the variable relations between possible acts and possible opinions. Indeed, some of his passages achieve such heights of abstraction that his thinking can no longer be related back to reality in its immediate sense; it advances, as it were, under its own steam, as if in an imaginary space, but those who follow it step by step can only find it utterly cogent. This is already the form of consciousness of pure mathematics—though without its contents—or of pure graph theory.

We may at this point agree with Oskar Morgenstern, who states, in his account of the mathematical theory of games and economic behavior, that "certain findings can no longer be conveyed in words"; a statement which he applies also to quantum mechanics.[13] We are thus transported into a scientific field of a wholly nonclassical nature, which admits no preconstituted notions as to the properties of the objects of discourse. The theory of games offers a good example of this. The model for discussing sociological or economic questions is no longer the primitive mechanical image of equilibrium, the cycle, etc., but rather games of strategy such as chess, where each player seeks to win, but controls only one set of variables, must include in his calculations the responses to his moves, and must try to predict his opponent's moves. Only the most basic elements of the theory are accessible to the layman, since it operates with such instruments as combinatorial calculus, set theory, matrices, and mathematical logic. I myself, being a layman, mention this case only because it constitutes a telling example of my thesis. The penetration of the experimental spirit into arts and sciences of every kind has as a necessary consequence their objects' loss of naturalness. The referents of the theory, being generated exclusively by the method one chooses, lend themselves to being analyzed and resynthesized in numerous ways.

Unavoidably this procedure thoroughly rationalizes the field of objects, which thus become more disembodied, more abstract, less intuitive, and finally—in a manner which defies easy description—"autonomous." Precise results can no longer be conveyed in words, and become clear only in the course of their derivation. Such methods have led to the most surprising results in the fields of natural science and technique. Intellectual history suggests the probability that in other fields the "classical" methods of, for instance, historiography or psychology will cease to be of interest for their most advanced students. That is to say, the frontiers of both scientific and artistic culture have become the preserves of virtuosos. Conversely, the popularity of, for example, Rilke, is held against him, without any properly aesthetic discussion of his merits; rather, sheer, immediate lyrical feeling as such is seen as stale, petty-bourgeois dross.

A Counter-Tendency: Primitivism

We have seen how artistic and scientific pursuits become steadily more abstract. This applies primarily, however, to those new ideas and approaches which, because they *are* new, are seen as "progressive" and "representative," since in contemporary culture whatever is newest always sets the norm. In all fields, the most admired are the innovators, those who blaze new trails and make discoveries, the revolutionaries. All the same, one might speak of a law of asynchronism, whereby the tempo of advance varies from one to another sector of culture, as well as within individual sectors. Within our culture as a whole, it is technique and natural science that exhibit the highest rate of change, whereas legislation proceeds more slowly, and criteria of social value and standards of prestige change more slowly yet. Similar phenomena are also found within individual fields. For example, the philological and historical sciences have not emancipated themselves quite as thoroughly from the intuitive as have sociology, economics, or indeed psychology, and maintain a more traditionalistic posture. Equally, by far the largest part of the reading public prefers Fontane, Rilke, or Hemingway to Joyce, Pound, or Benn; and most music-lovers remain unshakably faithful to the old masters.

It is interesting that cultural conservatism and *pragmatic* progressivism coexist in the same circles, those which never fail to pay attention to innovations in the technical-industrial fields. People in these circles increasingly view cultural pursuits in purely aesthetic terms, as possessing no material consequences or moral significance. In practical endeavors, however, a wholly different frame of reference obtains from that which applies to cultural pursuits. An endless diversity of subjective interests, of individual inclinations and personal variants, is expressed in and enriched by cultural offerings of the sciences and the arts. To increase the multiplicity and variety of stimuli, the most remote cultures are exhumed, and their remains displayed in great traveling exhibitions. But wholly independently of these experiences, men go on pursuing their political and economic interests with attitudes that do not reveal nearly the same degree of refinement and sophistication, but on the contrary are extremely concrete and down-to-earth. In comparison with that of today, the level of political discussion was astonishingly high a hundred years ago, when those involved in politics still believed in the moral fecundity of the sciences and in the possibility of a unitary, scientifically grounded view of the world. Now things are very different. Culture, in the old meaning of the term, makes no claim to bindingness and generates a kind of pseudoconservatism; new, "progressive" culture, if taken seriously, tends to atomize its adherents, and in extreme cases induces speechlessness. Meanwhile, social reality goes its own way.

The cultural mass media, such as radio and the cinema, are committed to a kind of primitivism which attracts much criticism but which, for economic reasons, is quite inescapable. For these industries must dispose of their products and do so as widely as possible, since they absorb considerable capital that cannot bear much risk. An artistically sensitive, competently made film might occasionally attain commercial success; but to produce one entails a level of risk usually found unacceptable. There are other ways to finance, say, a cultural radio program, but not a serious film that might fail at the box office.

The primitivism of which we are speaking results from the low standards the public expects in its entertainment; but there is also the "secondary primitivism" of cultivated groups, that is, a tendency to stun oneself with overdoses of gross stimuli. As is well known, the sculpture of black Africa and of Papua has influenced Western painting since before World War I, and early expressionism, in particular, sought extremely stark and shocking effects; compare Klee's oversensitive, transfigured dreams with Kirchner's strident masses of color. In fact, this neoprimitivism is, together with abstractness, among the factors that make contemporary art nonpopular.

There is a further variant of primitivism, which in certain cases has strong ideological roots: it consists in appealing to a notion of humanity as something formless and undifferentiated. As significant a proponent of modern architecture as Giedion, for instance, states: "We seem to find ourselves at the beginning of a new culture phase, in which man as such, naked, unclothed man, man unbounded by the determinations of social stratum, religion, or race, finds direct expression in forms and symbols which echo his inner sensations and affect his psyche."[14] Apparently modern architecture claims as its ideological partners the great metropolitan masses—historically unformed human beings, who perforce cannot find anything for them in the hermetic and inaccessible high culture of today.

An explicit return to the primitive can also be found in contemporary sculpture. Here, side by side with abstract constructs, sometimes of striking beauty, and with Calder's bizarre "mobiles" we find the brutal simplifications of Moore and Brancusi. In their lumpiness and roughness these remind one of works from the age of Constantine the Great; but they appear as if congested, lacking the starkness and impetus of ancient art. In his book on the age of Constantine,[15] Jacob Burckhardt tried to explain the phenomenon of the art of that age as the result of a degeneration of the race, of the appearance of men with truly "repellant traits"; but a few pages later he admitted that the deepest cause of the phenomenon could not be located or articulated. Today's comparable phenomena suggest the possibility that the primitivism in question was consciously pursued, that it constituted an elaborate style consciously seeking to express wider tendencies of the period.

Finally, we may use the term "primitivization" to characterize a further, striking phenomenon of contemporary artistic life: the cultural impoverishment of the verbal (not the mathematical) expression of thought. There is today a widespread lack of ability to convey meanings without voicing them—to comprehend shadings of meaning, allusive expressions, stylistic niceties, subtle conceptual distinctions. Everything must be expressed unequivocally and starkly. It is difficult for us to believe that at one point the works of Kant were actually read in cultured circles; and academic practitioners of the cultural sciences, in particular, often lament the number of students who cannot follow through a line of reasoning of more than elementary difficulty. Here, again, one should not content oneself too easily with sociological explanations; and we may perhaps agree with Burckhardt that the deeper causes of such phenomena cannot be located or articulated. Personally, I would seek such causes in a direction already indicated: the era of the Enlightenment is at an end, and with it the belief in the necessity of precise and abstract concepts. This imperils the feeling of the worth and intrinsic validity of science. The peril lies in the urge to simplify and visualize the knowable, and in the associated drive toward practical application; these impulses undermine the proud self-sufficiency of conceptual mastery and condemn it to irrelevance.

Diffusion of Technical Modes of Thought

In these first two chapters we have staked out the area with which we are concerned; we have also indicated some factors that influence man's inner life in our own time, and that come together to shape his evaluations, his interests, and his reasoning. Among these factors is the dualism of abstract conceptualization on the one hand, and primitivism on the other; the profound connection between the two is more easily intuited than described. But the central factor is the all-powerful superstructure where technique, industry, and natural science presuppose and complement one another.

Cultural research invariably shows that human consciousness is shaped by the culturally preferred modes of thinking and patterns of behavior. That consciousness is so dominated by the central themes of its own epoch that the views characteristic of its culture appear as the only natural and rational, or at any rate self-explanatory, ones. This holds true even today. One can easily show, for instance, that modes of thought developed in the technical context also impose themselves in nontechnical contexts to which they are inappropriate. The very fact that this needs pointing out indicates the depth of the *inner* transformation undergone by our ways of conceiving of reality. The very *structures of consciousness,* that is, the ways in which consciousness operates, change over long periods of time. To express this in philosophical terminology: the system of a priori representations of a culture can only to a secondary extent be derived from its contents; primarily that system inheres in the *forms,* in *how* reality is apprehended and interpreted.

It is possible to mention specific technical principles that have imposed themselves on a large scale in social and interpersonal relations. For example, the "principle of total requirement,"* that is, the avoidance of running idle, of dead weights, and of unexploited energies, has become a ground rule of the division of labor in undertakings of all kinds. If any individual is not "optimally utilized," a redistribution of tasks becomes mandatory. The "principle of prearranged effects,"† embodied in the rails of a railway system, becomes a ground rule of all planning; and the man with his hand on the button or the switch is the model of all duly operating management. The principles of

**Prinzip der vollen Beanspruchung.*

†*Prinzip der vorbereiteten Vollzüge.*

"standardized measurement" and of "interchangeable parts"* are well in evidence in the want ads, with their exact definitions of the required qualities and quantities (including age). The principle of "concentration upon effect"† has perhaps the widest application of all, from doctor's prescriptions to carefully planned propaganda campaigns, since the notion of optimal effects exercises a literally compulsive hold upon the men of a technical era. We are well pleased if an effect is reached with the most sparing means; and enthusiastic when, once "primed," such an effect keeps reproducing itself. Well-conducted propaganda, for instance, ought to influence people *automatically*—Biddles characterizes it vis-à-vis other forms of compulsion as the one which leads without arousing resistance.[16]

The widespread operation of originally technical principles in nontechnical territory throws doubt upon the popular notion that our own culture lacks a "style" of its own, although that notion might have seemed plausible at the time it arose under the influence of Nietzsche. The fields less adapted to the highly rational cast of our culture are no longer those where "principles" (*Gesinnungen*) are centrally significant; in fact, a pure orientation to principle is itself part of that cast. Rather, such fields are to be found where a calculable effect is intended but cannot be securely attained; for example, in the most modern forms of music, poetry, and painting. These indeed strive for the most direct, pure, and intense effect; but the probability of its being attained cannot be calculated exactly. For this very reason they must adopt a further, more reliable strategy for success: untiring propaganda.

**Prinzip der Normgrössen, Prinzip der auswechselbaren Teile.*
†Prinzip der Konzentration auf den Effekt.

Nonetheless, the search for sheer, irresistible effect is expressed in the manner itself in which work is conducted; it is, for instance, by limiting himself to giving the sparest, most minimal expression to a given pictorial idea, that Picasso (as we learn from Misia Sert) could manage to paint several pictures in a day.[17] Thus, he could produce a great deal more than could, for instance, Renoir, who, according to the same writer, had to spend one month on each of his seven or eight portraits of her, with three sittings a week, each of which lasted a whole day.

We mention this case because it seems to us of exemplary significance. Artists, scientists, etc., would operate rationally, in accordance with the spirit of the times, were they to cease striving in their works for a lasting, indeed a timeless, validity. It is in keeping with the context not only of a consumer society, but equally of a technical one where obsolescence has become a component of progress, that one should produce with an eye to speeding up turnover—as does Bernard Buffet, with his 2,000 paintings in ten years. The very impressions which the painters evoke are unstable; they constitute a mass of stimuli which decays, to use the language of physicists, according to a given "half-life." Since the effects themselves are so perishable, it would not make much sense to expend too much effort in producing them.

3 Social-Psychological Findings

Adaptations

WITHIN THE FRAMEWORK outlined so far we can show the nature of the transformations that have taken place. We shall adopt a sociological and social-psychological viewpoint.

The increase in the complexity of the structures of society, to whose secular advance industrialization has imparted such decisive impulse, has excluded great numbers of men from direct production and led to their urbanization. It has also imposed upon them functions so indirect, complicated, and specialized, that their moral and mental adaptation to their new situation has become highly problematical; in particular this applies to what one could call the maintenance of their social equilibrium. Not enough attention has been paid to the rapidity with which the concept of "adaptation" has come to be taken for granted in social-psychological research. The image conveyed by the concept is that of unalterable conditions of the external world, such that a given organism cannot escape them nor can it alter them; and indeed it is the overpowering hold of social, economic, and technical factors upon today's existence that has made the concept of adaptation uniquely useful in understanding people's behavior. The larger superstructures of contemporary civilization become autonomized and alienated (Hegel, Marx)—this makes the inner and outer behavior of individuals take the form of adaptation: a process whose advance is only in part willed and controlled, and to a much larger extent unconscious. This applies above all where the adaptation in question consists in a change in the modes of perception, in the thought patterns, indeed in the very structures of consciousness, rather than in affecting only the contents which one must receive and master. Primitivization of our patterns of thought through the spread of technical models, as described above, impinges upon consciousness, but does so as if autonomously, being neither chosen nor noticed.

There have already been attempts to place adaptation at the conceptual center of a whole system of psychology,[1] and only lately is skepticism being voiced against such attempts in the United States, where at one point they had begun to be taken for granted. David Riesman, in particular,[2] considers "wholly adjusted" as synonymous with "overadjusted," and "characterizes sheer social conformity as a defective type of conduct."[3]

The complexity of the superstructures of civilization is in fact so great, the social structures so opaque, that a specific discipline devoted to inquiring into them and describing them must come into being. This discipline is sociology, whose task consists in tracing the concrete premises and consequences of those structures, this measurement being necessary to all planning. An immediate consequence of all this for the individual is that there is no longer a close correspondence between his view of what he is doing, and his view of what happens to him: for instance, he dutifully attends to his job, and then he is deprived of it by a geographically remote crisis of which he has no conception. In these circumstances, he cannot react very differently from the member of a primitive tribe who, incapable of understanding why he has fallen ill, looks for a guilty party, who invariably turns out to be someone who on other grounds is looked upon with little favor.

It thus becomes more difficult to organize conduct rationally. Within the division of labor imposed by industrial culture everywhere (except in some sections of agriculture and in a few preindustrial crafts), all highly specialized activity is carried out at some remove from its ultimate product, and cannot control its own success or failure. Such activity easily becomes empty, sterile—and finally imaginary, when its counterproductiveness remains unnoticed. This applies above all where the higher decision-making functions within the economy, government, and administration have to be carried out without adequate knowledge of actual operational conditions, on the basis of defective or imprecise information, and where such defective information must also be used to determine whether any results at all have been attained. It also applies to the whole educational establishment, where the results of tests and examinations bear a very loose relationship to whether or not the establishment's goals have been met. "A professor can live in error and persevere in it all his life, he can destroy thousands, tens of thousands of intellects, and yet he stays on in his comfortable post and at the end draws a suitable pension. On the other hand, a farmer who has two crop failures in succession is a ruined man."[4]

This example may serve for many occupations of a highly indirect nature; specialists and functionaries of all kinds are often placed into positions where, in the absence of immediate and visible sanctions on how they think and operate, they find it difficult to learn from their mistakes. Yet, according to a well-known anthropologist, trial-and-error learning constitutes the earliest and most fundamental contribution ever made by cultural activities to the control of human behavior.[5] Never before (as happens frequently today) has the propaganda which must perforce accompany every kind of public activity suggested by its own bitterness and intolerance an awareness that it is destined to fail against the impalpable but insuperable resistance of men or of situations.

Everywhere one hears that the system makes specialization indispensable; and yet this very fact arouses much despairing opposition, particularly from top functionaries of all kinds, who are aware of how multisided their responsibilities are and of how unlikely it is becoming that they will find suitably unspecialized potential successors. There is a moral dimension to this problem; for only open-ended action, reinforced by its own successes, willing to take diverse risks and to assimilate past experiences and failures, can engage deeper and more personal levels of motivation, appeal to an individual's moral core. Yet the specialized tasks, made necessary in such great numbers by the industrial-bureaucratic system, do not require enough inner strength and capacity for growth to evoke a feeling of personal worth; they are no longer "open" on all sides, and therefore cannot constitute, as Dewey puts it, "experiences enjoyable in themselves."[6] Thus innumerable individuals are condemned to perform, rather than to live, their own social function, as is made clear by such expressions as the "occupier" or "holder" of a given post. The person is reduced to being the occupier or holder of qualifications, claims, characteristics, obligations, rights, etc.; that is, of abstract, categorical determinations. The most recent of these mindless neologisms is the expression *Sonderbedarfsträger* (in-charge-of-special-requirements).

An adaptation to spiritually meaningless, morally vacuous, and yet overpowering situations can take place in several ways: for instance, as opportunism, as a surrender to the changing circumstances. This is a plausible and frequent reaction, and those condemning it often fail to acknowledge the real factors bringing it about, and hugely overstate the significance of a principled posture. To maintain unshakable principles, to retain them in the face of varying and mostly unfavorable circumstances, until by chance some circumstances occur (entirely of their own accord) on which somehow those principles can make an effective impact—this orientation to existence recommends itself only for those rather few men who are suited for heroism or fanaticism. Another fairly frequent form of adaptation consists in taking a low profile, decreasing one's visibility, playing dead. A third and very significant one consists in what could be called feminization, meaning here that one emphasizes one's consumer orientation and a kind of passivity. We speak of feminization because hitherto it has been a privilege of women to engage in reckless consumption, and particularly in consumption of luxuries, with the best conscience in the world.

Now, however, this attitude is becoming generalized, owing to the extravagant production of consumer goods and the high average level of disposable income, with the attendant "obligation to consume," as Schelsky says. David Riesman's invaluable sociological discussion of the consumer culture takes as its points of reference advanced industrial society and a drop in the rate of population growth.[7] Mackenroth's inquiries, in particular, leave no doubt that the irresistible attraction of a comfortable standard of living and its related spending habits influences, by way of birth control practices, the demographic process itself.[8] For his part, Riesman clearly shows that the disposition to consume has become a controlling attitude, deeply affecting the relations between the sexes as well as sports and politics.

But the data so far available do not suffice to explain the attitude toward consumption. Neither the demographic stagnation of industrial societies, nor the greater availability of goods, nor the powerful suggestion exercised by conspicuous consumption would do as explanatory variables were it not (as Riesman has masterfully intuited and demonstrated) for the decline of "inner direction." That is, fewer and fewer people act on the basis of personal, internalized value orientations, on the basis of principles capable of stabilizing their total attitude in the face of randomly changing circumstances. But why are there fewer such people? Clearly because the economic, political, and social atmosphere has become hard to grasp intellectually, and hard to live up to morally, and because it changes at an accelerated pace. Reflect for a moment, for instance, on the shifts in the labels applied to an ethical value, loyalty, which can be realized only over long periods of time:

In the years 1936 to 1945 the term "loyalists" clearly designated, in England, France, Scandinavia and America, Spanish republicans of all shadings, including the communists, who fought against Franco. In 1945, influenced by Washington, Western governments withdrew their ambassadors from Madrid. Still in 1947 *Time* spoke of the famous "fifth column" of the Franquist General Mola as of one of the columns of mass betrayal. However, by 1954, at a time when treaties of alliance and agreements on military bases had been concluded with Franco, one of the crimes held against Robert Oppenheimer consisted in his having contributed so many dollars, between 1937 and 1939, to an American committee supporting the Spanish left; and an American well briefed on foreign affairs could refer as a matter of course to the supporters of Franco as "Spanish loyalists."[9]

In a world where such things go on, any belief in constant principles of orientation is in danger of being denied that minimum of external confirmation without which it cannot survive—unless, as in the case of religious belief, it appeals in the last analysis to an extraempirical guarantee, which by definition cannot be contested empirically.

On the basis of these premises one can understand certain phenomena characteristic of the contemporary spiritual and moral climate, in spite of their apparent contrasts with one another. In the context of the remarks made on abstract art, we suggested that contemporary, post-Enlightenment culture is more favorable to religion than was the preindustrial era of the eighteenth century. If there is reason to believe that we are entering a post-Enlightenment era, and thus one of cultural "crystallization," then this thesis would find support in the existence of a kind of "neodogmatism"; and indeed there are symptoms of the latter.

On the other hand one might equally well predict attempts to derive moral principles directly from the mass condition of contemporary society; to draw fundamental orientations from humanity conceived as an undifferentiated, basic condition. Such orientations would have to be strongly emotional in tone and lack conceptual precision, if they were to avoid being invalidated by the rapid change in the surrounding situations; and also in order adequately to play the role of a supranational and ecumenical morality of relations within a world much shrunken in size.

In fact these requirements are met by the current forms of humanitarianism, whose several manifestations range from the movement for Moral Rearmament to the world-wide diffusion of Human Relations studies. The ideal entertained by these various movements is that of a harmonious symbiosis whose central category is that of *acceptance*—that is, accepting each individual as he is, with all his characteristics, good or bad. It is impossible to do this without also accepting all the characteristics of the culture from which that individual has emerged; and to this extent the concept of "acceptance" can be considered as the germ of a world ethic which excludes a priori the moral and cultural primacy of European civilization.

In view of all these problems and difficulties (ultimately of a moral nature), one can understand, finally, the turn toward passivity. To complement this passivity come high levels of consumption, affording both physical and spiritual excitement by means of various "stimuli" and "experiences." The associated morality would equally tend to release tensions, to neutralize them, by barring the necessity of painful value choices and thus asserting its consistency with the "universal ethic" sketched above. An analysis along these lines can show that technique and economic arrangements, the arts and the sciences, morality and the content of consciousness of a given age all hang together; further, it can indicate that these phenomena, as if streaming out of the same source, all embody the same substance, out of which the twentieth century is constructed. Should one, on the contrary, try to separate causes and consequences, one would all too soon run out of meaningful statements. A mode of analysis in terms of causes and consequences does not lend itself to the discussion of developments on the largest scale.

The Loss of Direct Experience

Among the modes of adaptation to developments that escape intellectual or moral grasp by virtue of their very dimensions, one of the most significant is the contemporary loss of a feeling of reality. In keeping with a psychological law, desire correspondingly becomes unrestrained—another root of the prevalent disposition to keep quiet and keep consuming. It is not clear whether in the long run such acquiescence will be less damaging than a different kind of estrangement from the world, the kind that enhances fantasy, programmatic vision, the disposition to act. It is a sad sign, well fitting a topsy-turvy world, that sometimes one cannot deny some moral worth to Utopian fantasizing, insofar as it appeals to ideal needs which remain both unrenounceable and unrealized; whereas, by contrast, those who rationally engage in action are time and again shown up by the confusions they engender.

Already Sorel had observed that estrangement from everyday existence most enhances the ability to live in an imaginary world. Everyday existence as understood here involves the pressure of variously changing situations, in which those considerable elements which are familiar and predictable become mixed with surprising and unpredictable ones. But after all, the industrial culture shares with the magic-ritualistic cultures of primitive peoples the property of severing man from everyday existence. Between early primitivism, which spins its resilient net of fantastic accounts and rituals over the world of experience, and the later primitivism, engendered by the unbearable pressure of the monotony of a narrow sector of experience, there is not much difference from the standpoint of world estrangement. In one case the horizon is obscured by myths, in the other by newspapers.

But "experience" is definitely one of those things to which the elegant Hegelian notion of a transformation of quantity into quality applies. For, as we know, it is through the many-sided exchange of immediate and concrete experiences that stable certainties become formed; but also, vice-versa, truths which otherwise would merely "float on the mind," can be brought down to earth only if they match a wide range of concrete reality. Our industrial, administrative, and scientific specialists, operating as they do in a rarified atmosphere, can only vaguely and distantly control the consequences of their actions, which often are fairly difficult to visualize. Such contexts constitute an excellent feeding-ground for extravagant fantasizings, where, unless one opts for consumerist acquiescence, the undernourished social instincts can find some expression. If an individual feels that he is but a replaceable and slightly worn cog in a great machine; if he becomes (justifiably) convinced that the machine can run without him, and he comes into contact with the consequences of his actions only by means of statistics, or graphs, or in the form of his paycheck—then of course his sense of responsibility decreases at the same rate as his feeling of helplessness increases. Only the routes we have mentioned remain open to an individual whose personality has been radically crippled.

In the realm of economics, too, we are confronted with the same phenomena, and particularly with that of consumer passivity. Wilhelm Röpke writes that

> an excess of division of labor makes man's vital energies wither. . . . Every day modern man becomes less capable of attending to his own needs. Convenience foods replace those prepared at home; ready-made clothes make dispensable the traditional housewife's labor of cutting-and-sewing; music from records and

the radio replaces the playing of music at home; the car and the soccer game take over from the personal practice of sports. Finally people receive ready-made views and opinions from those thought-machines, the press, the radio, and the movies. Considering that in some cities in the United States more illegitimate children are sought for adoption than are available, it seems that there already are people who want even their own babies made by others.[10]

Some consequences of this are significant from the social-psychological viewpoint. The situation compels individuals "to form opinions and feelings concerning aspects of reality far beyond their intellectual and emotional reach."[11] People are continually being pressured to form and express this or that good sentiment. Already more than a century ago the philosopher Arnold Ruge spoke of the "haze of sentiment" in which those to whom it is not given "to scale the heights of the spirit in knowledge and vision" seek refuge and compensation.[12] To be quite fair, one must admit that it has probably never been more difficult to gain a solid knowledge of the broader picture. In fact one can only speak of *knowing* as a component of conscious and purposeful action; in a true sense, we only know what lies within the reach of our lived experience. It is true that, beyond that experience, we can get to know much by learning, yet we then assume that we are being taught something that reflects the experience of the teacher.

Now, to acquire such grounded knowledge about the givens of the social, political, and economic world of today, and in particular about the connections between those givens, is possible (if at all) only for those who enjoy some opportunity for initiative, who hold responsible positions in those larger contexts and are in a position to evoke a response even from yet-uncertain facts. But there are very few such people; all others cannot possibly "scale the heights of . . . knowledge"; and yet the consequences of what happens in those wider contexts pursue them into their very homes, and demand of them some kind of reaction.

At this point, not many possibilities are open. One can react on the basis of mental association and of feeling, that is in a primitive fashion; or, equally primitively, one can personify the circumstances, and adjust in a more or less compliant or rebellious fashion to "what those on top do." It is fortunate that sometimes there is some justification for relying upon the honesty and knowledgeability of those holding positions of responsibility; but primitive reactions prevail, at least in the public realm. If one considers the narrow formulas with which people seek to make some sense of the enormously complex connections between historical events, one cannot avoid the impression that these are stereotyped rituals, which lack, however, the unvanquished optimism of old-time magic.

In close connection with these facts stands the trait of our times analyzed by Ortega y Gasset, to the effect that

> . . . Today's average man has the most mathematical "ideas" on all that happens or ought to happen in the universe. Hence he has lost the use of his hearing. Why should he listen if he has within him all that is necessary? There is no reason now for listening, but rather for judging, pronouncing, deciding. There is no question concerning public life, in which he does not intervene, blind and deaf as he is, imposing his "opinions."[13]

This remark seems to us to fit best those educated people who themselves qualify as mass men by virtue of the fact that they react exactly in this fashion. In any case, thinking on the basis of sentiments, associations, and impulses is easily employed when large, obscure questions of public significance are concerned. Here the same individual who cannot count on his own ability to make sure of tomorrow's bread seems to have a magical formula that accounts for everything and solves all problems.

Opinion as Second-Hand Experience

Since special organizations have come into existence for the study of public opinion, the formation of opinions has been receiving some attention. It should be considered as a special form of "order-setting" (*Ordnungsstiftung*). Man possesses the ability to set up very numerous, diverse, and quite precise basic models (categories); given his risk-laden condition, he also possesses a deeply felt need to interpret the ill-ordered confusion of the world as experienced and of the flow of events, in such a way as to yield a maximum of order, coherence, and uniformity. In our previous discussion of the particular fascination exercised by automatisms, we argued for the existence of a human need for environmental stability. We have also offered practical and theoretical examples of particularly significant stabilizing devices, such as the habitualization of the circle of action, the rhythmic periodic rotation of machines, and astrological doctrines; the latter, unprovable as they are from a rational viewpoint, must owe their large following to the peculiar attraction of the sense of order they impose.

As Hofstätter has convincingly shown, superstition—to which *all* men are prone—should be seen as but a special case of the general tendency to overestimate the level of ordering within the stream of events.[14] A superstitious person simplifies the world by means of coincidence formulas, or (pseudo) rules of event sequence; he interprets the course of the world as more orderly, simple, and interested in man's welfare than it is in fact. Given the joint occurrence of two unusual events, say the first H-bomb explosion and a particularly cool and wet summer, it is nearly impossible *not* to see there a fatal causal connection, to be possibly avoided in the future. We thus encounter a typical "stereotype" of public opinion, since the view in question is very popularly held.

This stereotype is only distinguished in its logical form (the imputation of causal connection) from many others which take the form of simple judgments. White Americans, for instance, view "the Negro" as religious-minded, superstitious, lazy, and musical: the degree of simplification, the gain in order, and the related credibility confer upon such formulas a degree of acceptance which could never be justified by the "kernel of truth" they contain (as do all such prejudices). In Germany, for instance, there is a widespread prejudice to the effect that Rommel was our best general in World War II. The experts competent on this question are in fact the military specialists from the other side; and they put forward other names, particularly that of Manstein, but wide German circles still stick to their preference for Rommel. Rommel's performance, however considerable, could not have generated such fame were it not for the influence of romanticism about Africa, and Goebbels' efforts to throw a heroic light upon an ill-fated campaign. The above American stereotypes about Negroes can be used to account for Father Divine, the well-known Negro religious leader, as well as for jazz music or indolent black servants.

We have barely begun to inquire into how such schematic and yet precise representations arise. Naturally a significant role is played by the generalization of individual experiences. Further, the activity of image formation might be explicable in terms of the Gestalt psychology of perception: that is, the tendency for our psychic equipment to "touch up" all manner of irregularities, gaps, deficiencies, and obscurities in the immediate givens of experience, and to favor self-enclosed, coherent, and easily apprehended forms. In addition to such mechanisms, our feelings and affects clearly have their own influence on our opinions, and this usually favors simplification, and particularly polarization. If our passions make us see everything in black and white, ultimately this is due to the fact that passions are virtual actions, or at any rate emergent commitments of the will; as such, they always tend to project stark alternatives.

All such arrangements favor simplification, that is, they overstate the orderliness of reality. Given the human being's high susceptibility to stimuli, and the constraint under which he finds himself to engage in action, such a tendency is clearly life-sustaining. Superstitious as they may be, schematic representations of states of affairs or of causal connections still possess some "facilitating" significance, for they preserve us from the tension of doubt, uncertainty, and hesitation. Also, the intensity with which they are entertained dissuades us from exposing them to actual experience, which might easily disconfirm them; exactly to that extent, they allow us to make a definite commitment to a given position.

Considering all these obstacles to rational and objectively grounded judgments, it is astonishing that man is not more estranged from and blind to the world than he is. In fact, we still have to mention a further, significant factor in the formation of our opinions and convictions; namely, the mediation of experience itself.* Between the individual, whose range of experience is always very narrow, and the vast, imposing processes arising from the social, economic, and political superstructures, there necessarily intervenes a third component: "second-hand experience." That which was once learned through "hearsay" is today communicated by the information industry, by the press, the radio, etc., although these sources have not altogether replaced those resting on interpersonal contacts, such as gossip and "the grapevine," which in turn often draw upon information supplied by the mass media. Many of the facts in circulation originate from the activity of large organizations involved in collecting, diffusing, and therefore interpreting, facts.

The influences which thus affect "news" and "facts" include the technical necessity of brevity; the organization's rules for assessing the significance of the information; and the ultimately unavoidable subjectivity of the individuals in charge (as well as wider considerations). Such mass media organizations do not operate in a vacuum but under the influence of other centers, in turn possessed of their own tendencies. Naturally such organizations are rational enough to consider to whom the information will eventually flow, and to what effect.

Since the processing of facts takes place within organizations, it is unavoidably a directed process, for organizations *qua* organizations must operate rationally, that is, purposefully; and the purpose in question is exposed to the influence of multiple factors. At least equally significant and unavoidable is that deformation of news which follows quite unintentionally from the fact that nowadays

**Das Mittelbarwerden der Erfahrung selbst.*

the "meaning" of an event is not contained in or determined by the event itself. Thus the shell of the fact, in order to become news, must be filled with the opinions of the commentator.

What we call opinion, then, is formed by the sedimentation of all these processes. It is unavoidable that opinions be formed in this manner, since first-hand experience must perforce be inadequate, and its deficiencies must be filled in by schematic contents, at any rate when the questions are significant enough and there is enough pressure for a position to be taken on them. It is impossible to do without such opinions, since within today's immense world of facts one is thrown back upon secondary sources, which offer themselves to us, in print and in image, with varying degrees of reliability. One only needs opinion to acquire a *bienfaisante certitude* within such a sea of uncertainty.

On the other hand, it would be simplistic and misleading to regard opinions as just nebulous, half-tenable, deformed representations of facts concerning which one might, with due effort, acquire adequate knowledge. This may sometimes be the case even in fields where the lack of adequate information can have very considerable repercussions, for instance as in the case of political elections. Thus it is difficult to see why in June 1954, 65 percent of women should answer "I don't know," or give vague or wrong responses to the question, "What is a 'governmental coalition'?"; or why for many years between 32 percent and 40 percent of the respondents have stated that the salaries paid to ministers are "what costs most money to the state."[15] A serious information campaign beginning at the level of elementary school might well eliminate this kind of ignorance. But very different significance attaches to opinions held on matters on which no objective factual knowledge can be obtained, and where at the same time there is an "irresistible" pressure for opinions to be formed. One such matter is that of the "character images" which nations cannot help forming about one another, as they enter into closer and more frequent contacts. For instance, the self-image stereotypically entertained by Americans differs considerably from the stereotypes which the French, the German, the English, and others, entertain about Americans; but, as Hofstätter remarks in his important discussion of this problem, we simply possess no data on the actual distribution of character traits in any nation, ethnic group, or religious collectivity.[16] There is no need to emphasize the political significance which may come to attach to such public but objectively ungrounded opinions concerning other peoples, races, etc.

Finally, more than mere logical interest attaches to those cases where the state of affairs is intrinsically ambiguous, and becomes defined only by means of opinions and particularly of evaluations. If discrepant evaluations on such matters keep arising, the state of affairs itself remains unresolved and forever clothed in obscurity. This applies, for instance, to many of the dramatic events discussed by Margret Boveri in her extraordinary, valuable, and solid book on treason.[17] It is often apparent how the evaluations of the actors, which constitute an integral part of their action, and the evaluations of those directly or indirectly affected, which in some cases concur in determining the quality of the action, constitute an intrinsically ambiguous event, of the kind with which our logic cannot adequately grapple. Hence any discussion of such situations which tries to "stick to the facts" runs into ever larger and less soluble problems, such as those raised by the question of when Hitler's regime became illegitimate.

One should particularly notice that just as in the case of individuals so in that of collectivities, the necessity to act often exceeds the available motivational resources. In such cases one finds oneself compelled to form opinions in order to motivate those actions which the situation forces upon one, since ultimately one cannot act as an automaton. In the last months of World War II one could well see that many members of the army were perfectly aware of the impending German catastrophe. Yet, the closer one was to the front, and the more the enemy's pressure was felt, there was simply no way one could act according to that awareness; one indeed had to soldier on but could equip himself with the appropriate belief, that in "secret weapons"—a belief which in fact became widely accepted exclusively for that reason. In situations full of drama and tension, it is not possible to infer what conduct will follow from an opinion, however frequently expressed (such as, in this case, the acknowledgment that the military situation was untenable). Conversely, public opinion overestimates the assurance with which one may predict conduct from the opinions or convictions held. Even opinions expressed with the strongest conviction are yet in no way virtual actions. This much is certain; but it constitutes another stereotype to hold the opposite.

Experience and Ethos

It has thus become possible to advance more than sufficient reasons for the loss of direct experience described above. The loss or attenuation of the sense of reality, considered as the very core of all our difficulties by as acute an observer as Schumpeter,[18] found expression in the arts much earlier than in psychology or the social sciences. The resoluteness with which all arts, over the last fifty years or so, have turned their backs upon preconstituted reality should be considered as a kind of ontological vote of no confidence, though most artists are incapable of replacing the reality they reject with anything nonsubjective. However, when Musil, in his great novel *Der Mann ohne Eigenschaften* (The Man without Qualities), lets destinies and events unfold in the context of an action "parallel" to something nonexistent; or when Kafka keeps introducing dream motifs into descriptions couched in a pedantic-naturalistic style (as later on the surrealist painters were to do), then the doubts about reality projected by these great artists attain a kind of *objective* significance.

In order to perform its distinctive mission of interpreting such doubts instead of merely expressing them, philosophy must seek the support of sociology and social psychology. And here again one must refer to matters previously discussed. Industrial development has filled the world with a vast array of organizations, all connected with one another by innumerable functional interdependencies. Large-scale planning goes on not only in the spatial, but also in the temporal sense; yet most plans get changed before they are carried out. The economic, political, and social atmosphere prevailing between the continents presages stormy upsets whose effects will be felt in every home and every heart, but it has become utterly impossible to attain an adequate, rational knowledge of the impending developments. The compulsion to form opinions about opaque, but dramatically significant, facts might perhaps be considered as a form of facilitation, except that it remains unclear how far the convictions held or expressed can impinge upon the actual courses of action.

The widespread feeling of malaise; the occasional spasms and fevers of the social body; the constant stream of ideologies, hastily constructed and as hastily taken apart; the elimination of all attempts to cover up egoism, or the artificiality and thoughtlessness of the excuses offered for it; the suddenness with which ghastly brutality erupts—all these phenomena make it blatantly evident that mankind has not yet established a sound moral relationship with industrial culture, or, even more certainly, *all* of our current environment. Röpke suggests (mostly on account of the political implications of a mass civilization) that the development of our psychomoral foundations lags behind the advance of the division of labor (or social differentiation).[19]

All these phenomena demand to be interpreted in a comprehensive fashion. In view of the fact that currently not only European, but also serious American literature "imparts no sensation other than that of a loss of the sense of reality,"[20] it is no longer feasible to account for the phenomena in question by referring to the national and historical particularities of different peoples. At bottom, the crisis appears to us not as a religious one (as is often asserted) but as a total one, in the sense that the basic coordinates themselves of the interpretation of the world have become doubtful. In any case no provision is made for a fundamental human need, which must be fulfilled if man is to enjoy any security: the stabilization of the living space* eludes technical culture just as much as does the stabilization of social space. As Schumpeter says, "creative destruction" lies at the very core of industrial culture.

In anthropological terms, this entails that our convictions, our obligations, even our opinions lack a solid point of reference. We do not possess an invariant reservoir of

* *Stabilisierung des Lebensraumes.*

usages, habits, institutions, symbols, ideals—the "cultural real estate" against which we may confidently measure our conduct. On the contrary, we can never afford to relax the tensed-up, awake state of our consciousness; we must maintain a chronic state of alarm, continually monitor our circumstances, continually ethically reorient our conduct, continually articulate and revise our basic commitments. And we must do so on a stage where foreground and backdrop, characters and lines, are constantly changing. Under such conditions it is difficult to generate that minimum of pressure to conform which ultimately no society can exist without. We are left with a variety of practical, theoretical, moral, and existential viewpoints which can no longer provide a basis for mutual understanding. That such a view is not exaggerated can be shown by considering any public issue of any significance, from that of conscientious objection to the Suez Canal crisis. In each case we find not just a conformation of incompatible basic commitments, but an inability to agree even on the strictly factual aspects of the issues, on the question of what the issue is all about.

Since all the novelties which assail us involve some break with tradition, even honest resolves lack inner conviction: for traditions form the very basis of the individual's own essence and will. Without them, one can keep restlessly active all over the place, and yet lack any inner sense that all this busyness carries any moral significance. In changing the world, men have demolished the invisible supports of their own spiritual identity.

Perforce the need for ideology increases by way of compensation. A "well-arranged ideology"[21] can to some extent make up for the lack of order and coherence in reality itself, and also plays an important role in connection with "second-hand experience." Poorly assimilated data, barely suspected connections, facts, and effects which loom on the horizon—we cannot master these things intellectually but we can at least put them into ideological boxes.

Many of the symptoms discussed here, from the demonic dance of modern art to consumer acquiescence, indicate a less and less secure hold upon reality, and this also gives an idea of their moral significance. For it is impossible for man to order spiritually his image of the world and his view of existence unless he possesses some kind of unconscious confidence, which in turn depends entirely upon the continuity and tenability of his actions.

Yet for obvious reasons modern culture is not easily amenable to moral evaluation. In the first place, economic and political power relations become depersonalized, and at the same time more and more significant. Thus the ancient conceptions of justice and mercy, protection and obedience, piety and loyalty, anger and devotion, all notions of a semifamilial nature which used to supply immediate terms of reference to morality, lose their relevance. In their place, the imagery that becomes more and more fitting envisages devices which, if they fail to operate satisfactorily, may be replaced by others. As Max Weber has emphasized, every personal relation, even that to one's slave, can be ethically regulated; but not an economic system which, thanks to commercialization, has become a self-enclosed, self-ruling entity, whose autonomy is only seemingly curbed by the advent of planning of production and consumption. Furthermore the autonomy of a thoroughly industrialized economy is particularly striking in the diversity of the effects it produces—effects so inescapable and pervasive that they reach even the birth rate—this autonomy appears suprapersonal to an astonishing extent, nearly "metahuman."

Equally, scientific research does not lend itself to moral evaluation; the researcher himself is deprived of his moral autonomy, since he neither sets the problem nor "decides" (as the laymen think) the uses to which his findings are put. The problem is generated by the existent state of knowledge, and the logic of experimental procedure entails that exact cognition already comprises the mastery of effects. Thus there is no question of deciding how to apply the knowledge gained; this comes of itself, the object of research imposes it upon the researcher. The cognitive process is, itself, a technical process. The interface between science, technical application, and industrial exploitation has long since become the province of a distinctive superstructure, which in turn is automatized and ethically indifferent. A radical change in this situation is only possible at one of the two ends of the process: at the point of departure, in the will to know; or at the point of arrival, in the will to consume. At either end, the change would require an *ascetic* attitude of such novelty as to signal the advent of a new era.

The moral aspect of modern culture raises very interesting problems. A thorough politicization of morals could be seen as an emergency response, a simplification, an artificial limb compensating for some deep functional deficiencies. If there are no longer any external limits to the tasks we set ourselves—as seems implied by our current interest in space travel—then perhaps such limits could originate internally, from the fact that ultimately our senses set the boundaries within which we are capable of *reliable* moral responses, and thus limit their reach to proximate realities. This would be a counterpart to a previous statement, to the effect that one can feel sure only concerning the premises or consequences of action, or at any rate the components of action and its *controlled* effects. Beyond the visible and surveyable proximity encompassing all "real contact," both intellectual and moral, it is possible for calculation and conceptual elaboration to take over without disconnecting the controls exercised by convictions. But in entering this realm we leave behind what we can represent to ourselves freshly and literally; thus we cannot expect to discover within it a set of rules, controls, or constraints.

In his last book, Bergson worked out a similar conception of the natural boundaries of a stably organized social unit, using the notions of "natural" and "closed" society.[22] In this context he suggests that if statesmen of the appropriate quality are so hard to find, this is because we expect our politicians to solve at one stroke, in all their complexities, problems which the very size attained by society had previously made insoluble. Aristotle raises similar issues when inquiring into the optimum size of a state, which makes sense in view of the issue central to ancient political philosophy, the issue of stability. With this in mind Aristotle had stressed the significance of the comprehensibility of affairs of state.

Today one might formulate the same problem as follows: what are the psychological effects of the destabilization of conduct due to the dimensions of our situation, in particular when our goals (and our destiny) go far beyond the narrow range of our innate human resources? When concepts can no longer be "filled out" with intuitions, values with tangible attainments, situations with instinctive awareness—in these conditions the collective soul must turn to the imaginary, the fantastic, and even the unreal. Since each of us, at one time or another, stands at the edges of these unknown areas, each will fill the huge gap between what he does and what happens to him with pseudorational fantasies, and with extravagant, overstimulated feelings that can no longer find expression in concrete actions. Such fantasies must be morally simple, since even in the moral realm it is the contact with empirical reality that induces complexity, engenders differentiation. Thus arise monocultures of slogans and "principles"; one need merely look at a few pamphlets and periodicals or attend a few shows, to be aware of the rate at which arid imaginations think up concepts, and these are then introduced in one way or another into this or that controversy, without it ever becoming clear where they have any bearing upon reality, or which reality they bear upon.

One can also detect a tendency to take up crystallized, simplified, dogmatized systems of thought, each capable, for all their differences, of providing their supporters with preconstituted positions upon all moral and philosophical questions. Complex and sophisticated sets of ethical or intellectual issues become politicized and stereotyped. On the one hand this has the facilitating effect of yielding simpler orderings; on the other it induces a narrowing and contracting of the horizons of thought. What is remote and intrinsically complex is brought within reach, until it appears easy to grasp and to control—in the same way that eyeglasses can create a kind of secondary proximity. A reality that is infinitely complex, many-sided, and changeable is thus made artificially simple and graspable so as to bring about the proximity indispensable to all morality; and all this happens in the name of a superpersonal and unconscious necessity.

4 The New Subjectivism

Changes in the Modes of Experience*

ALONG WITH THE transformations discussed so far, others have been taking place within the narrow realm of private existence. Interpersonal relations are becoming simpler, and losing the fullness of structure and of meaning they possessed, their diverse norms and authoritative audiences. Over the centuries the forms of public life had seen a fusion between a feeling of purposefulness on the one hand, and a keen sense of taste on the other; as this fusion had gained extension into the private realm, those forms had served, there, to protect the individual from himself. As in the case of language, they also provided a support for those who could not do them justice, and whom they surpassed both in wisdom and in effectiveness.

The dissolution of such forms has taken place in stages. They proved unable to withstand the tyranny of the economy: the emphasis on profitable management of assets economically undermined those very circles which previously had both perfected the standards of taste and possessed the power to make them binding on others. As later the economy itself, internally weakened, had to seek the support of the state, practically all public life became tied up with the state, and the latter in turn became primarily

* The expression *Erlebnis*, here translated simply as "experience," conveys lived, "internal" experience, as contrasted with the expression *Erfahrung*, emphasized in the preceding chapter. See "Relations between Internal and External Experience," below.

an administrative and welfare state, committed to "planning."[1]

Any individual transplanted into our own times from the vigorously concrete cultures of antiquity, of the Middle Ages, or even of the baroque era, would find most astonishing the conditions of physical proximity, and the lack of structure and form, in which the people of our time are forced to vegetate; and would wonder at the elusiveness and abstractness of our institutions, which are mostly "immaterial states of affairs." Because of this we seem to possess practically no patterns of conduct whereby men can exist *with one another;* even sport has turned into a show supplying stimuli to passive masses. The family remains as the sole "symbiotic" social form, and owes its stability to this monopoly position, a stability astonishing in so changing a culture. The family appears as the true counterpart to the public realm, as a refuge of privacy. As public life loses its deeper, symbolic import, as institutions devolve into statutes and statutes into traffic regulations, the private sphere separates itself wholly from this context. However, it becomes a sphere wholly given over to immediacy, where individuals interact in the whole range of their strengths and weaknesses; their lack of mutual distance multiplies tenfold the conflicts, and family members must use the scanty reserves of their accidental, personal qualities to moderate these conflicts.

This situation accounts for the astonishing contemporary diffusion of psychological awareness, as well as for the unprecedented differentiation, explicitness, directness, and lack of reserve of individual psychological states, whether overt or covert. What a difference from the times of Molière, when the overt display of personal, intimate features immediately qualified a character as comical in the eyes of an audience with sensitive antennas, and with serious restraints upon its own modes of expression.

Our proximate world, where individuals develop random peculiarities of character under the pressures of the complexities of modern existence and their own diverse circumstances, and where those peculiarities mirror each other, thereby both displaying and generating individual psychic physiognomies—this world supplies the themes of that uniquely Western art form, the psychological novel. The latter conveys the sociological situation described above, and shows quite clearly that with the decay of solid social orderings is associated the development not just of psychology, but *of the psyche itself*.

An apposite precedent to this phenomenon is the "New Comedy,"* which had for late antiquity the same significance as the novel has for us; as soon as the *polis* "ceased to lay down the law about everything, the way opened up for the emergence of previously unthinkable stirrings of the psyche."[2] "All instincts which no longer find an external outlet, turn inward: hence the growth within man of what was later to be called 'soul,'" writes Nietzsche.[3] This is somewhat overstated, but in psychology one cannot do without some such overstatement.

As we have already argued elsewhere,[4] one cannot seriously doubt that modern subjectivism is rooted in the wider cultural situation; in the presence of a flood of stimuli that overtax our capacity for emotional response, an emphasis on inner elaboration and "psychologization" represents an attempt to keep things under control; external determinants are vital, though unacknowledged. Emotional reactions can no longer be invested in an external

* An artistic movement in Hellenistic Greece, having as its main figure the playwright Menander (343–291 B.C.).

world which has become so reified and deprived of symbolic undertones; raw nature no longer opposes us with a felt resistance, the exertion of bodily effort has become largely dispensable. How could this not have as a consequence a continuous flow of internal experiences, which must be monitored through unceasing awareness and reflection? It is here that art, law, even religion become subjectivized and weakened. "Ideas" bud forth everywhere, and one can only deal with them by discussion, this being the appropriate form of external elaboration. This intellectualization and subjectivization of a culture screened away from action is a novelty of our own historical era; it is a component of the very air we breathe. Those who cannot see this must be refusing to look.

According to this conception, for which we have laid an extensive groundwork elsewhere, the immediacy with which the single psyche perceives itself and in which it lives constitutes nonetheless an indicator of the outlying general conditions, social and cultural. It is not a matter of the self-perception being straightforwardly "caused" by the general conditions, but of an "if–then" relationship obtaining between the two. We believe, in fact, that what distinguishes the inner existence of an individual in other times and cultures from that in our own is not so much the difference in contents that fill consciousness now and in other times, but rather the difference in formal and structural qualities and modalities of lived experience. Yet it goes against the grain of our own time to admit such a mutation in the very structures of consciousness, rather than simply in its contents; such an admission contrasts with the prevalent contemporary self-complacency (which exists side by side with a penchant for self-criticism). However, some past thinkers had already grasped the nature of this difference; for instance, Emerson, as shown in a passage from his *Essays* which Nietzsche excerpted:[5]

> The costly charm of ancient tragedy, and indeed of all the old literature, is, that the persons speak simply—speak as persons who have great good sense without knowing it, before yet the reflective habit has become the predominant habit of the mind.

Here Emerson contends, perhaps rightly, that the ancients lacked the chronic reflection and cautious deliberation which characterize us.

Let us try to grasp the logic of the relationship between inner and outer experience. The lack of stable institutions, which at bottom are nothing but preformed and customary decisions, makes heavy demands upon man's ability and willingness to deliberate; and, by demolishing the bulwarks of habit, exposes him defenseless to the casual flow of stimuli. These stimuli give rise to interests, gains, and needs which attain a more or less stable equilibrium within the individual. The characters thus formed seemingly possess qualities of their own, but only insofar as they have not yet found an occasion to shed them. Only these and similar social-psychological premises can account for certain easily observed phenomena—for instance, for the pressing need to communicate with others, which is absent in sound and well-established personalities, and largely explains our proclaimed individualism; and for our sophisticated understanding of the psychological features of others, dictated by our keen feeling of the threat others may represent for us, and unnecessary where the encounters of individuals are controlled by all manner of tacit understandings, based in turn on the validity of well-established conventions. The same considerations may also account for the fact that frequently, as if in response to the necessity continually to probe, reflect, postpone, ponder, one artfully assumes a posture of equanimity and detachment, combined in fact with a considerable degree of watchfulness.

Finally, the already discussed willingness to exhibit to others one's most intimate concerns, which accompanies the abandonment of traditional forms of discipline, and is presupposed in turn by the sophistication of contemporary psychology, can be placed in relation to some distinctive features of our time: in particular to our overreflectiveness to the habitual self-consciousness of our inner existence. Our external conduct is a nonimmediate, often tortuous outcome of our internal states, and exactly for this reason must be treated as a symptom and subjected to interpretation; correspondingly, there is a relative lack of direct, naive, general emotionality. No exact proof of this is possible; but a reading of autobiographies from previous centuries compellingly suggests that the emotional sphere has since undergone a deep change. Explosions of primitive, naive, unfeigned, often criminal passion seem to have occurred much more frequently in the past than they do now. We may assume that the framework of consciousness characterizing contemporary, more-or-less urbanized man, with its abstractness and its reliance upon second-hand experiences is also associated with a comparable emotional modification. More and more what one sees in operation are emotional schemata, empty formulas capable of being filled with almost any content, "shells of emotion."

In other words, sentiments also become "second-hand." See for example the pin-up girl reproduced in millions of copies, as a schematic trigger for equally schematic erotic reactions. Pictures of this kind equal political propaganda in their ability to evoke second-hand emotions. These emotions can in turn be easily expressed, lying as they do near the very surface of consciousness; for this reason they can be easily diffused through modern advertising techniques; and in fact they also penetrate the realm of moral feelings. One cannot account otherwise for the widespread presumption that what one wants is in the general interest, and that one's subjective emotions somehow represent requirements of "human dignity." One often comes upon the form of moral confusion (not identified as such because widely taken for granted) that consists in feeling responsible for "the West" but not for one's city or village; for "culture" but not for one's bookshelf; or even for "religion."

Elsewhere we have shown that it is entirely understandable and necessary for emotions to become schematized.[6] At any given time the psychic realm possesses a certain number of *topoi,* of patterns of feeling, of sentiment, of what is ideal and intellectually valid; and it is by becoming thus formalized that such contents of experience acquire stability. Our psychic existence receives back the influence of our social conduct, which in turn is regulated and controlled by obligations, which correspondingly select and make obligatory a number of sets of motives and emotions. As Blondel says, "there is much that is artificial and conventional about social sentiments"; and he adds that this can also be said of aesthetic, moral, and religious sentiments.[7]

The truth of this is a basic tenet of cultural anthropology, to be taken into account at all times; but in the present its validity is particularly in evidence. The media of second-hand experience to which we are exposed must have recourse to special tactical devices in order to traverse our overtaxed brains and our exhausted sensory apparatuses.

For instance, in order to evoke the attraction of readers, listeners, or viewers, news reports must develop a pistol-shot style, compressed and dramatic. The news must be juicy, the headline provocative, the picture sensational, or it will not overcome the apathy of the overstimulated. Feelings and passions responding to such stimuli naturally possess to a high degree the previously described characteristic of being like empty shells or play tokens: contentless but striking.

Actually this dramatization, originally the stylistic form of news reportage, is becoming more and more widespread, no matter how earnest and honorable its intentions. Whenever one must appeal to the public on behalf of a given object, dramatization necessarily takes place, with all it entails: slogans, superlatives, vivid imagery, appeals to a sense of danger or of threat.

Relationships between Internal and External Experience

These derivative, generalized, and schematized states of feeling, and these similarly formed opinions—all are within individuals capable of a degree of sophisticated subjectivity unknown in previous centuries. An internal existence that has turned in upon itself can reach a further degree of advance; there already are many individuals in whom a thought or a feeling not only operates in itself as an actual process, but goes on to constitute a stimulus to further thoughts or feelings. The decay of ideals and values which echoes within the individual the transformations of the new era, produces a sophisticated awareness of psychic phenomena. As we have already stated, the modern psychological novel mirrors this development, by focusing on states of feeling and analyzing with an unprecedented awareness their subtlest niceties, and recording their interaction in the minutest detail. Sentiments which previously were unproblematically lived through now acquire a literary and thus a public existence. Furthermore, the large-scale reprinting of records and legacies of other times and other cultures further enriches one's psychical existence by allowing it to appropriate that of remote other beings through empathic understanding (*Verstehen*). Never before were so many people equipped with such fine sensors. Seen from this viewpoint, the era of the masses appears as one in which the most extravagant ventures of subjectivity are successful in claiming public recognition and attention.

Our argument, briefly restated, is that the vast, complex, and artificial structures of our technical civilization have engendered identifiable psychical reactions, two of which deserve emphasis. The first can be described as a tendency to lose contact with reality: the increasing mastery over matter seems accompanied by an increasing estrangement from the world, by the development of fantastic and extravagant programmatic ideas, convictions, and group feelings. For instance, over the last few decades leading politicians from all sides, none excepted, have made spectacular mistakes in their evaluations of conditions and possibilities, in spite of the wealth of relevant information available to them. The extremity of their errors is inexplicable unless we realize that those who feed on ideologies unavoidably misperceive and understate the options open to them for action, and this not because of personal incapacity or unwillingness, but as a result of an objective process.

The second point is that the modern *psyche* develops

at the same time as the *science* of it and as the *art* which mirrors it. The evidence of the dissolution of society is provided not only by contemporary science and art (as one might say such evidence was provided by the psychology of Plato and of Aristotle, or by the New Comedy), but by the distinctively modern psychic states *themselves:* in particular by the extent to which they become self-conscious and articulate, by their ability to react upon themselves, to feed upon themselves, to become differentiated.

Many readers will object to this last assertion, since they are unable to see how those aspects of one's internal existence that one considers unproblematical and views as "natural" can in fact be connected with external data of a supraindividual nature. Yet one must learn to accept the existence of such connections. As early as 1921 in a work on Klee, one of the first modern painters, the critic Wilhelm Hausenstein stated:

> The subjective element does not stand above all others. However, in this awkward moment, it is the only one that counts. . . . When any objective element fails . . . so does any natural feeling of commonality. And when the latter disappears, the subject becomes the only law and the sole concern. Our era rejects an objective point of reference. Within it . . . it is unthinkable that art should appear otherwise than as in Klee's graphic art, whose boundaries are only set by the extravagant reach of his subjectivity.[8]

Shortly before, Hausenstein had drawn up a kind of balance sheet of the ruinations, mendacities, and profanations that had afflicted the façade of everyday existence in the years of World War I and immediately after, and concluded it with the remarkable words, "the comprehensibility of things and people has sunk even below the threshold which it had previously reached." The author connects with this slippery reality Klee's thematic subjectivism; he makes it comprehensible that in the face of such a situation a sensitive, anguished soul should take refuge within himself, and that the "protest against technique" should play a role in this inward flight. "This is more or less what the world still has to offer: scattered patches of objective reality; residues of things disrupted by the passage of a fast train or a car, discolored, flattened and fragmented by the movie camera." It is as if our thesis concerning the loss of contact with reality and the origins of subjectivism in the introversion and sophistication of the atomized psyche had been worked out in the first place from a reading of modern art, and then generalized. In fact, this is not what happened; rather, our point of departure was the observation that around the middle of the eighteenth century there simultaneously began European industrialization, the science of psychology, and sentimental-psychological literature such as Richardson's novels (or *Werther* in Germany). Since then, states of consciousness which originally appeared as utterly sensational novelties have become "functionalized" and have crept into everyday experience. It is highly improbable that that triple simultaneous occurrence was just a random coincidence. What at first was grasped and articulated only by the most outstanding minds, is now seen by everyone: that we live in an external world constructed industrially, thoroughly technicalized, harboring millions of ego-centered, self-conscious individuals, all seeking to enrich their own psychic existence. None of these individuals questions that a momentary, irresponsible quickening of that existence by means of any stimuli and experiences whatever should have become a continuous concern; indeed, this is taken wholly for granted.

To see that this is true one need only admit what intellectual history (*Geistesgeschichte*) has long since claimed: that the "leading spirits"—the trail-blazers—address concerns and states of reality and formulate structures of experience that later will be held to be "natural," in the sense that they could not possibly be otherwise. The great psychologists of the eighteenth century, as for instance Hume, came very near the views we entertain today on the connections between the cultural environment and the states of the psyche. By means of the "psychology of mental associations" they thought they could show that psychic existence follows the same laws as nature. They erred mainly in their conception of the laws of nature, which do not that much resemble those of classical mechanics. Yet the latter are now the laws central to the functioning of technique and of the industrial world; and once this is understood, Hume and the classic French works of psychology can be seen to have argued—correctly—that in order to comprehend psychic states and processes it is necessary to project them onto the external world.

Fictions and Games

Let us return to the idea that the ancient New Comedy represented for classical antiquity the same thing as the psychological novel does for us. As the *polis* lost its central position, says Howald, previously unthinkable psychical phenomena made their appearance. The old *polis* embodied the harsh demands of the struggle for existence, the necessity for vigilance and discipline, risk and mortal danger. Once "Alexander's vigorous domination" (Hegel) lifted these burdens from the Athenians, the New Comedy entered the stage with its new, lighter mentality. Our own situation is analogous. The struggle for existence becomes less harsh; the strain and the working discipline imposed by low preindustrial productivity become relaxed; the workday can become limited to an extent that would have struck our grandparents as sinfully short and accommodating; the welfare state activates mechanisms for the redistribution of wealth. At the same time, subjectivism reigns undisturbed, and in the wealthiest continent a complex psychological theory—psychoanalysis—acquires the standing of a world view.

Earnest, or playful; tragic, or comic: no third term exists. Where is the boundary between the humorous and the comical, the playful and the childish, the artistic and the artful? Many of the highest artistic and philosophical renderings of the modern psyche, in fact, have something about them that is artificial, forced, and contrived, because of their extreme sophistication, subjectivism, and freedom from the pressure of necessity. You need only to distance yourself and consider them soberly to detect a touch of frivolity, a low-keyed eccentricity. In their originality and vigor they approach the level of genius, but cannot quite attain it because they lack the naiveté of true intellectual inventiveness, and seem instead to mix sophistication with something that is raw and overly spontaneous. "All veils of the heart have been torn away. The ancients would never have made their own souls into the subject of fiction to this extent," says Madame de Staël.[9]

The notion of the "subject of fiction" can refer, as did Madame de Staël, to the writer's need to communicate, his urge to replicate his own experiences, tensions, and reflections in a novel or a philosophy. To some extent, however, one can refer it also to "the soul" itself. Even in the absence of any neurotic disposition, today the high degree of awareness, the moral arbitrariness, the lack of intrinsically authoritative models, and the oversupply of ideological themes—all these things yield combinations of highly doubtful authenticity, which still find a wide audience. Everywhere we observe a need to validate one's worth, a need historically unique in its intensity and in the amount of insecurity it generates; this need lies at the base of all manner of artifices and pretenses, which often expose one to ridicule.

Yet one should not forget that this apparently omnipresent emphasis on one's own worth is but the individual reflection of a sociological condition: the nonexistence of a taken-for-granted system of social rank. Such a system would eliminate the problem of validating oneself, by imparting some objective validity to everyone's position, which would thus cease to be a subjective problem—another example of how changes in external givens become transformed into immediate, direct affects and psychic impulses. What produces an "inferiority complex," for instance, is not a comparison with objectively given standards of valid conduct leading to the realization that one does not measure up to them. Only at a time when religion enjoyed an uncontested spiritual dominance could Geulinx write: "Humility is divided into two parts: respect of self, and contempt of self."[10] Today it is impossible to identify standards of value valid for everybody, and this is what engenders the inferiority complex—the social role someone plays is *no longer* unquestionably accepted as valid by some other group. For this reason a society as complex and opaque as ours raises in the most acute and compelling fashion the problem of personal worth, and thus generates innumerable opportunities for people to put forward extravagant claims, develop compensatory attachments, and suffer from handicaps. Thus, insofar as the present is characterized, as Finer suggests, by profound individual perplexity and insecurity with respect to spiritual values, unavoidably the individual psyche will constitute a "subject of fiction"[11] of considerable significance.

More people than one might think are aware of the threatening nature of this situation, with the result that quite often in their personal relations individuals play together at not taking themselves too seriously. This phenomenon is part of a wider set of circumstances, which Ortega y Gasset and Huizinga have previously discussed, and which again constitutes a parallel with the ancient New Comedy: "A far-reaching contamination of play and serious activity has taken place."[12] On the one hand serious matters are dealt with in the framework of play, or even of the game of chance; on the other hand, childish concerns are taken very seriously.

In Huizinga's famous book on play there is a chapter on puerilism, which lists some symptoms of childish behavior: the need for banal entertainment, a need easily met but never satiated; the search for gross sensations, for mass ostentations, etc. As he rightly says, on a psychologically deeper level there is a lack of humor: people react in exaggerated fashion to each other's words, impute to one another bad intentions or motives, show intolerance toward opinions other than theirs, are easily taken in by any delusion flattering to their sense of self or to the group they belong to. "The puerile habits I have in mind are, in themselves, as old as the world; the difference lies in the place they now occupy in our civilization and the brutality with which they manifest themselves."[13]

Thus, the earnest and the playful often exchange their ancient provinces. Sport becomes the refuge of national resentments, and its increasing commercialization is accompanied by increased fanaticism, animosity, trickery, and media partiality, as well as by repeated excesses of ill-feeling. On the other hand, politics loses the emotional resonance it possessed for our fathers and grandfathers; it becomes almost a form of entertainment, and it will not be long before elections turn into a kind of championship finals for the masses. For example, in the German electoral campaign of 1955, cabaret shows, fashion shows, and open-air film shows first came into use, since it had become "impossible or difficult to get good attendances at old-style party rallies."[14]

It seems that only the presence of visible, determined contrasts guarantees that something will be taken seriously, as in the case of economic and occupational life, in sports, and in those contexts in which serious matters of political power are settled. Elsewhere things slip easily into farce—as do even the most tragic ancient themes when a contemporary playwright such as Anouilh or the cinema takes them up. On the other hand, it causes some surprise to a social psychologist to discover how utterly humorless are the proponents of modern art, in spite of its unmistakable components of the playful, the grotesque, and the absurd. Hazard a pleasantry or two on modern art, and you will find yourself confronted with openly expressed, poisonous hatred. This is strange, since in this case one cannot see that there is a massive and threatening opposition in operation.

Both from the social-psychological and the historical point of view, what makes these phenomena remarkable is the difficulty of finding closely analogous developments in other situations. The well-known saying "bread and circuses" brings to mind crowds of shiftless southern Europeans in the late Roman Empire; and this imagery does not fit our own situation, since zeal, involvement, and ambition are still dominant aspects of our society (although for some time experienced teachers have been complaining of the puzzling, growing manifestations of passivity, intellectual laziness, and contemplativeness among our young people). Also Huizinga's "puerilism" applies only loosely to the frequent playful aspects of public affairs, for what is involved is not the puerile variety of nonseriousness and it lacks the inspired and committed quality of children's play. However, the light admixture of farce we have in mind does account for some significant innovations. For instance, it allowed Benn to use a jargon of Berlinese inspiration—throw-away, sarcastic, and trenchant—in order to handle significant contemporary situations and ideological questions, without violating thereby his audience's sense of what is and what isn't "possible."[15]

The problem thus posed is probably not among those one can properly "solve," for in such matters knowledge consists in the first place in describing phenomena and in seeking meaningful connections between them. It then becomes possible, so to speak, to encircle the theme by bringing to bear upon it several categories, none sufficient by itself. Riesman, for instance, describes in a most imaginative and compelling way contemporary man, who operates under the decisive influence of others and also slowly adopts a consumer's attitude in cultural matters. Yet the notion of "consumption" does not quite convey the components of nonseriousness, illusoriness, and lack of conviction to which Benn reacts so sensitively. What is involved is probably a kind of "facilitating effect," a paradoxical phenomenon associated with civilization, something like a loss of feeling for the specific weight of reality. Modern painters have developed devices that render this very effectively. Similarly, to use a not precisely fitting analogy, in its time the "romantic irony" conveyed the loss of the sense for significance that the intellectuals of the time had suffered as an ultimate consequence of their continuous traffic with such concepts as "infinity" and the "absolute spirit."

The clear current turn toward an interest in poetry and in literature dealing with the visual arts of all times appears symptomatic in this context. Apparently the search is on for something serious, in small dosages for individual absorption. Over the last decades, significant novels have taken large-scale problems as their favorite themes, on which to make valid and significant statements which were prevalently ironic, satirical, Utopian, tongue-in-cheek, or otherwise detached: exemplars range from Kafka's forlorn hero K. to Thomas Mann's redoubtable genius Dr. Faustus. Should this trend consolidate itself, it would mean taking leave of a form of literature which for decades had dominated the international scene, and which could be considered as "psychology objectively, irony subjectively." At the same time one would be confronted with a maximum of estrangement from the world and distance from reality, insofar as nonironic attitudes would find expression primarily in the realm of minor art. Much of recent painting would seem to belong to that realm, to the extent that, according to a forgotten but very insightful statement of Hausenstein's such painting is essentially a matter of drawing, and in that sense constitutes graphic art, "for very little support is being offered to the pictoric in its full extent."[16]

Much may be gained, if in the future the violated Muses opt for concealment, fleeing outside the reach of advertisement and commercial interests, and art worthy of its name goes on to live an esoteric, little-known, private existence. It may thus avoid the fate of Picasso, whose seventy-fifth birthday was put to advertising use by a "well-established brandy producer" with the slogan, "Are you for or against Picasso?" and whose world fame inspired a tie shop in Düsseldorf to focus its front-window exhibits, "not without good taste," on his lithographs. This makes a fine parallel to the betting firms making book on local elections, and a realistic social-psychological description cannot afford to miss the symptomatic significance of such phenomena.

It seems that many of these current phenomena take on farcical features only because one gets their dimensions all wrong; scaled down to a fitting size, several times smaller, they would not give offense to such an extent. Freud's theory applied well to dreams and neuroses and on this scale it was an outstanding success, but it ceased to convince when applied to religion. And concerning modern art, Worringer has delivered himself of a judgment as clever as it is skeptical: "Do we not, in dealing with current art, use words way above the possibilities today open to it? . . . As far as art goes, we have lived for quite a time in a system of paper money, and continue to behave as if this money were still backed by gold."[17]

5 The Secular Horizon

Agrarian Morals and Industrial Morals

WE MUST NOW consider some moral aspects of our theme, and focus upon certain convictions which, while themselves the product of historical development, are so caught up with our whole existence that we have ceased to be even aware of them. In this context we shall presuppose, without further discussion, that besides those traits which are nearly universally considered as moral virtues, there are other *mainsprings of conduct* which underlie moral attribution (*Zurechnung*) and stand at the source of our feelings of obligation. The discussion will address those which, while culture-bound, can be considered as basic axioms (whether seen as decisions, convictions, or as particularly compelling attitudes). We are dealing, in fact, not so much with consciously articulated convictions, as with semi-instinctive orientations, drives operating on a mass scale. These manifest themselves in the individual in the form of strong, unselfconscious determinations, and are brought about by radical changes in the conditions of existence, as well as by the mutations irresistibly wrought in the structure of consciousness by science and technique, by the industrial and urban environment. These determinants are so compelling that even those who challenge this truth contradict themselves in the very language they employ.

Descartes had already given expression to one of these determinants in his famous formulation, "*maîtres et pos-*

sesseurs de la nature"[1]—in this he predicted the imperialistic control of nature which characterizes modern science. Science has imprisoned and interrogated the gigantic forces of inorganic nature, while technique has sentenced them to forced labor. Descartes had expressed in its purest form what Maritain has called the "anthropocentric optimism of thought,"[2] and had allied that optimism with the boundless pride that constructs whole worlds from a few principles. Descartes enjoyed such pride by virtue of his threefold private capacity as a nobleman, a genius, and an independent spirit. Yet the same pride proved as capable of popular diffusion as his whole philosophy, and proceeded to conquer all regions of reality, taking them apart, rearranging them, drawing new boundaries between them.

One cannot overestimate the accompanying transformations of the human soul. The history of culture probably knows only two truly decisive watersheds: the prehistoric transition from a hunting to a settled culture, and the modern transition to industrialism. In both cases we are confronted with a total intellectual and moral revolution. The transition from hunting to cattle raising and agriculture must have required many centuries, and the associated stresses must have been of the greatest magnitude.[3] For it by no means involved only a transformation of economic life, but rather a total restructuring of all attitudes, a restructuring that left nothing unaffected. The gods, for instance, had to change from metamorphic demons and animal figures into something resembling the human form, with ties to the locality. The Paleolithic hunting culture must have possessed very little in the way of mythology, which must be largely of Neolithic origin. Settlement entailed previously unthinkable possibilities for the ordering of families, and for kinship and other groupings. Increased populations, differences in wealth, authority relations of an unprecedented kind, all emerged and opened up new kinds of risks, obligations, freedoms, rights, and constraints. Although we can be sure that such a transition was accompanied by endless crises, these have left no direct traces, and our own forebears saw the resultant new foundations of human culture as possessing eternal validity. In fact, from the Neolithic era to the beginnings of modernity, agriculture remained the economic basis of humanity, and there was a moral implication of that agricultural basis: the care of animals and the cultivation of plants always involve a reciprocal service—as they exist for man, so man exists for them. To combine the objective, stable, and overlapping features of the nurture of man and of the propagation of animals and plants, to take nature's task of maintaining and fostering life as one's own—these were momentous developments, which we believe to have taken place not primarily through trial and error, experimentation and reflection, but rather by way of side-effects, as secondary results of *cultic* behavior, which in turn derived from extremely archaic animal rituals handed down from the hunting culture.[4]

Fast-growing agrarian societies become dependent upon atmospheric, climatic, and vegetative elements whose laws they have not yet mastered; and this chronic and unavoidable dependency reaches into the very core of the sense of existence. Most of the population lives, settled on the land, in conditions which make it necessary to attach connotations of service and duty to labor and in which the old animal cults no longer offer conceptions of divinity adequate to the floods and droughts which sometimes threaten whole peoples. Above all, since all economic activity is connected with living matter, social, ethical, and economic categories have no reason to grow apart. Also, the attitudes with which man best copes with his dependency upon unpredictable natural events are those that include a basic willingness to undergo sacrifices. It is unthinkable for him to rebel against the notion that he too is a creature; at any rate he lacks the superstitious credence in the omnipotence of men which characterizes his urban counterpart. Only in the cities is one inclined to argue, when this or that thing proves unsatisfactory or fails, that since man can do anything "if he puts himself to it," the trouble must be the work of ill-intentioned people, who must be identified if one does not yet know who they are.

The agricultural epoch (which embraces almost all of history) was characterized by a whole series of experiences of the greatest significance. The notion of capital, of a property which gains in value and demands to be put to use, already appears within the agrarian economy (in Sumerian, *mas* means both "interest" and "newborn animal"). Here capital formation is, as it were, an ontological process, a reality within the very substance of life in the world, with its economic significance and its ethical legitimation. Thus there can arise no generalized doubt concerning the rights of property, but at most specific objections concerning specific rights. The sacredness of private property constitutes one of the key features of agrarian societies: it delineates the sphere of one's will and power, and the sphere of one's moral responsibility for the prosperity of living things. This total acceptance of the notion of property, as well as the hankering toward stability, the thinking in terms of seasons, years, and generations, and finally the willingness to submit to something conceived as general and as outside one's influence (a conception not yet found in hunting cultures, and no longer in industrial ones), all these basic categories are crystallized into the central concept of traditional culture, that of *jural order* (*Rechtsordnung*). "The jural order is the decisive political component in man's transition to a peasant way of life," writes Heichelheim.[5]

Machines working upon dead matter; fashions that change every couple of months; the pressure of transitory conditions, of chance conjunctures and of deadlines; and finally the belief that by changing some premises of social existence the world's suffering could be brought to an end—these basic elements in a postagrarian culture unavoidably impinge most deeply upon the jural order and upon property, loosening up their intimate connection. "The British upper class has been bled almost to death in the most gentle manner, and with a measure of cooperation from a conscious victim."[6] This cooperation is an example of a decisive political component of the advanced industrial era. Such reforms have successfully availed themselves of the respect which for centuries has attached to legal precepts. But how long can one count on the habit of *group discipline* leading the victim itself to assist in its own execution, if Tocqueville was right in his prediction:

> I see an innumerable multitude of men, alike and equal, constantly circling around in pursuit of the petty and banal pleasures with which they glut their souls. Each one of them, withdrawn into himself, is almost unaware of the fate of the rest. Mankind, for him, consists in his children and his personal friends. As for the rest of his fellow citizens, they are near enough, but he does not notice them.[7]

What does Tocqueville depict here? Did he foresee the cities teeming with millions in the welfare states of the rich affluent societies? Did he address the conditions that prevail when all of politics comes to revolve around vast arrangements for the administration of everyday existence? Or did his frightening vision of an "orderly, gentle, peaceful slavery" likely to become "established even under the shadow of the sovereignty of the people" point not to a political entity, but rather to a consumers' dictatorship, which converts itself into a sensation of freedom?

If one is not to raise insoluble moral problems, and wants to avoid the danger of an excessive confidence in one's judgment (which even as great a man as Tocqueville did not always avoid), it is best to consider such characteristics as symptoms of possible future developments. Unless we are misreading the signs, people have begun to aspire to stability again, even though they have at first expressed this aspiration through a search for narrow contexts and petty satisfactions. Both the moral well-being of the individual and his mental health depend decisively on the solidity of jural order and on the integrity of juridically sanctioned institutions. For every legal institution is oriented to some measure of reciprocity, and to that extent necessarily connects obligations with advantages and compensations. A commitment to the service of others becomes durable and dependable only if it assigns some advantage to those undertaking it; and in fact it is on this basic pattern that institutions such as marriage and indeed all those entailing mutual obligations are constructed. The relationship between egoism and altruism would seem irrational within the individual, but becomes rational and coherent when externalized in the form of a legal institution. Put in another way, man's ideal and egoistic interests never converge in the individual, or rather do so only to the extent that they correspond with those of others *in the external world.* He who exclusively pursues an ideal impulse, inspired by merely subjective stimuli, makes himself into a fool. For when the institutions surrounding the individual are in a process of change or are being dismantled, leaving him to fend for himself in a vacuum, he can only act rationally by being egocentric.

For this reason institutions whose legal validity rests on unquestioned assumptions have a most significant influence upon an individual's internal constitution. They relieve him of the often taxing search for an appropriate line of conduct, by preselecting and clearly designating the line to follow, and by rewarding compliant conduct with status and economic advantages, or with the gratifying consciousness of having done the right thing; at the very least, they do not burden the compliant individual with disadvantages. Under these conditions morality is not thankless, it is not a thing which yields no returns or leaves one at a disadvantage for his pains. Nor is it a laborious process, a matter of many unconnected decisions. On the contrary, morality becomes a lived, established habit, supported by the ideals as well as the interests of others.

The quotation from Tocqueville contains another point of interest. It seems to foreshadow those smaller associations and those face-to-face groupings which arise throughout mass society and increasingly attract the attention of sociologists.[8] "The age of the masses is also the age of the small, special groupings, of relationships of trust to which one makes a real, active commitment, of groups which enlist like-minded associates."[9] Here the individual's isolation is moderated; for this reason, such informal, mostly nonpublic groups acquire increasing significance, and constitute the topic of a whole field within American social psychology. Man's inhumanity, as Tocqueville describes it, does not manifest itself within circles of young contemporaries or friends. All these particular smaller bonds, taken together, cement the larger social edifice. However conspicuous in the foreground, the large-scale, formal organizations, with their pigeon-holed individuals, are not the whole picture.

Upheaval and Asceticism

We have shown that the transition to industrialism dislocates some foundations of order and morality which had become established over millennia. From the psycho-moral viewpoint, the decisive fact is that one cannot assume an ethical posture toward inorganic nature (coal, electricity, and atomic energy). Thus, *the conception that there are boundaries to the permissible means is not an intrinsic datum of the very process of production,* which accordingly no longer imposes it upon everyday existence. The objectified cosmos of the industrial system of production and exchange, as we have already seen, is not directly accessible to ethical demands, which lose significance in its context. From this standpoint, there is no difference between the superstructures of a "private" and those of a "planned" economy. In each case, when dealing with inorganic nature and the conversion into labor of its powers, only the posture of the *maître et possesseur* appears plausible. It is already implicit in the basic structure of the whole domain, i.e., the experiment: what is known at all about the processes of nature is known by means of the experiment, which delivers its results directly into our hands. A theoretical mastery of the process is possible only through the practical mastery of it.

In knowing and utilizing inorganic nature, there are in principle no ethical, but only technical constraints upon the tasks one can set oneself; such boundaries, accordingly, are always provisional. In itself, the object of the undertaking in no way hinders the expansion of the intention to master it. Today, we live in a world where the trend is toward a maximum of 15 percent of the total population remaining engaged in agriculture, and therefore in direct contact with living matter. Thus, the tendencies indicated are bound to penetrate very deeply into the psycho-moral make-up of the civilization as a whole. Little wonder, then, that an indeterminate feeling of fear is becoming so pervasive. What men fear are not the monstrous destructive energies of the atomic nucleus, but their own; not the H-bomb, but themselves. They sense, rightly, that they cannot count on an internal constraint upon the use of a power one holds in one's hands to emerge suddenly in the final state of a development, whose main tendency for two hundred years has been exactly to remove such constraints, to foster and enhance a purely objective, rational, and technical concern with effectiveness.

If one is to evaluate two "commanding needs" (Nietzsche) which we are about to discuss, and which characterize the current social-psychological situation, one must reflect that the foundations of our spiritual constitution were laid by two hundred years of Enlightenment; and that even irrationalistic countertendencies, as for instance existentialism, operate fully within the terrain originally staked out by the Enlightenment. That movement is now past its prime; the era which it had brought into being is at an end. Yet it has left in us deep, unconscious traces; for instance, the modern psyche's extravagant penchant for communication is a product of the Enlightenment culture, and would have been unthinkable in the sixteenth century. The same applies to another common aspect of the characteristic modern mental categories, a tendency toward reflexivity which we take for granted to the extent of not even being aware of it. Anybody who, for instance, speaks of "religiosity" instead of "God," by the same token thinks within the Enlightenment tradition.

The Enlightenment era appears to us at an end; its premises are dead. Yet its consequences are still with us, in particular a deeply rooted tendency to consider a certain number of things as obvious. We repeat, the Enlightenment premises are dead. For instance, today we are not as likely to believe that reason, equally present in all men, can from its own resources attain nontrivial knowledge; or that nature is "reasonable," i.e., knowable through and through, and incapable of contradicting the criteria posited by an educated and right-minded humanity. And today, when the bourgeoisie of the whole world often seems secretly to consider itself beaten, it is worth recalling that at one point it considered as utterly certain that what it considered rational would ultimately triumph in history.

Such dogmas and axioms are so much things of the past that it has become difficult to even recollect them; at the same time, traces of that era are still indelibly inscribed within our psyches. In particular, the conviction of the "omnipotence of reason," originating with the Enlightenment, is no longer expressed in those terms. Yet, we would argue, that same belief, that same conviction, persists in our industrial era in the form of a boundless, optimistic willingness to set targets, to lay plans, to organize and reorganize. The characteristic mode of operation of the modern spirit consists in attacking the foundations, manipulating the central components, reconsidering the originating assumptions of whatever one is dealing with. Hans Freyer recently characterized this trait as a tendency to "make and remake" things.[10] The generalized, almost routine use of the term "revolution" in all spheres indicates something quite meaningful; everywhere foundations are attacked in order to rearrange them. This applies to the arts, to poetry, and to the scientific world view, as much as to politics itself.

The notion of reforming this or that flawed aspect of an otherwise valid order, rather than undertaking a radical shake-up of that order, holds little appeal for most people. Margaret Mead, for instance, sets for sociologists the task of planning the rebuilding of the world, and announces a firm resolve to invent a new social order.[11] Thus the concept of reconstructing the bases of society is not clothed, as in previous revolutions, in the guise of a struggle around specific rights. Instead, it is put forward quite specifically in the form of a "plan," of a deliberate organizational operation. This tendency to transform the social order itself and not just this or that aspect of it, made its first appearance about a century ago, and at that time people correctly saw it as metapolitical. Consider, for instance, a statement by the Seine prefect of police, in a report written in 1847: "This tendency for anarchic parties to leave aside properly political questions [!] and to commit themselves to views of a new social order . . . is more active than ever, and the government should keep it under constant watch."[12] The same expression, that one is not dealing with *passions politiques proprement dites,* was used in Tocqueville's speech to the French Chamber of Deputies of January 29, 1848 in connection with something of much greater magnitude. In Germany Lorenz von Stein made the same remark in 1843. What is involved, James Burnham remarks, is that people think precisely along the lines of a manager or an engineer intent upon organizing a modern industrial plant.[13]

This optimistic belief in planning is the contemporary

form of belief in reason, and as such has something instinctive, emotional, and uncontrolled about it. It represents a faith in the inner validity of our modes of thinking, and in their ability to master even the most large-scale and most complex conditions, those which characterize whole societies; and being a faith it is not rationally grounded. In fact, as can every true faith, it can dispense with any foundation.

One can only assume as a matter of belief and hope that such an infinitely complex reality will conform to those modes of thought, should one try to implement them, and that in the process nothing unexpected will occur. Reason cannot err, it speaks the language of reality, and reality cannot contradict it: again, these were basic assumptions of the Enlightenment. Ermatinger writes that the concept of the autonomy of reason, and the dissemination of that concept, had consequences of immeasurable historical significance, greater even than those of the Reformation.[14] Most other premises of the Enlightenment have since died and been forgotten, in particular the axioms of man's goodness, of the immanent purposefulness of the process of nature, of the universality of moral impulses. What remains, however, is the confidence of rational knowledge and thought in their own competence, in the sufficiency of their own powers. Only the arts of our own century have thrown doubt on this faith, and have thereby placed themselves into a polemical relation to the rest of our culture, with its imperturbable rationality, its mathematical-technical triumphs, its social edifices and balances. In fact, within the babel of voices, through which the contemporary arts try to attract attention, the voice of anxiety remains dominant.

The rationalist tradition has not only endured, it has

developed at an astonishing pace, particularly under the impact of both World Wars. What only thirty years ago appeared as Utopia, confronts us today as almost an accomplished fact. As late as 1922 Tönnies still considered as a Utopia "the creation [!] of a social condition . . . where the development of society, which so far has taken place in a spontaneous fashion, is—to use the Hegelian expression—placed upon its head, that is upon reason. The program would be: abolition of all profit-oriented private property, regulation of production according to common needs and the needs of the individual, and normative regulation of needs."[15] Such a program no longer strikes many Europeans as Utopian; even those who reject it must argue their rejection, and do not simply dismiss it as absurd. Such is the speed of the advance of general "rationalization" (Max Weber).

The second basic assumption concerns the question: To what uses should the power acquired over natural forces be put? The unanimous, unquestioned response is: to raising the standard of living. Huizinga was right to emphasize that as the technical capacity to master nature attains its peak, so also does the aspiration toward worldly comfort and worldly goods.[16] Finer states that "in our own time, the heavy emphasis of all contenders is upon the acquisition of wealth. Most men would like to be richer and work less and in more agreeable occupations."[17] The truth or falsity of this statement constitutes a social-psychological question of the greatest significance.

Much speaks for the truth of this view and especially the astonishing absence, in our time, of any ascetic ideals. This must seem strange to the historically informed, since in all previous times marked by a sharp advance in the demand for luxuries, ascetic ideals always existed as a counterweight to that advance, and were never basically challenged. The individual who renounced the goods of this earth always enjoyed a moral authority, whereas today he would be thought mentally defective. By asceticism we mean here any voluntary renunciation of the gratification of consuming goods, however motivated, and at whatever level of need, including the highest (the pursuit of merely aesthetic or cultural interests, or even of the pleasure of expressing oneself in the generally accepted, public phraseology).

Concretely, asceticism operates to strengthen the inner capacity for affective experience; it adds to the integration and solidity of the personality, and at the same time it sharpens the social impulses and increases spiritual awareness—it is, as it were, a concentrating upon oneself which makes oneself more available to others.* Under primitive conditions, the same effect of inhibition and discipline, which asceticism spontaneously induces in the individual, is enforced by the hardships of the external environment, and for this reason asceticism is rather rare in primitive societies. On the other hand, in the modern era, in spite of many shattering experiences to the contrary, we share an image of man as an intrinsically innocuous being; and this image blocks any insight into the most intimate and dangerous connection in the human soul, that between well-being and cruelty. Yet such an insight had been attained by Jung-Stilling, of whom it is said that he painted frightful pictures of the future, expressing his sorrowful conviction that religionless Christians would reveal the greatest aptitude for all manner of atrocities, because of the level of comfort they had attained.[18] This is a

Ein Sichentäussern gerade durch Sichkonzentrieren.

troubling statement, in spite of its failure to indicate exactly to whom it refers; and it makes a deep impression in view of the fact that in the twentieth century it was not the Asians who took the lead in atrocities.

As the Enlightenment's belief in reason has currently become formalized into a general readiness to plan and reorganize, so the other discovery of the Enlightenment, the vindication of human happiness, lies at the root of another need of industrial society—the need to consume. The right to well-being is an assumption as basic and unchallenged as the need for the reorganization of society, which indeed is ultimately the means to that end. The urgency with which the masses aspire to both ideals flows directly from the Enlightenment tradition, in the form newly imparted to it by the changes in the conditions of existence and in spiritual certainties associated with industrialization. This urgency also constitutes an adaptation of deeply rooted instincts to the resultant new situation; for the industrial system has always acted to raise the living standards of the masses, and by the same token to put its opponents, those of the conservative-traditionalist strata, with their distinctive intellectual and economic reserves and restraints, on the defensive.

Almost all industrial products, from electric light to silk stockings to radios, were at one point luxury goods, but thanks to mass production became objects of mass demand. There is little doubt that, when it operates without interference according to its own spirit, industrial production is not oriented to traditional, stereotyped needs, but takes a role in generating the demand for products which it first develops on its own initiative.

This process is irreversible. One cannot but seek to provide growing populations, whose expectations continually rise, with increasing quantities of goods. Yet one also ought to be aware of the intellectual and moral costs of the process. The system not only presupposes the right to well-being; it also tends to exclude its contrary, that is, the *right to renounce well-being,* particularly insofar as the system itself *produces and automatizes the need to consume.* Here, perhaps, lies the root of all currently dominant forms of unfreedom. Descartes's notion of man as *maître et possesseur de la nature* presupposes a "nature" around us so malleable and fruitful that it would not impose on us any necessary renunciation of its bounty. By implication the social order itself was presupposed not to make such renunciations obligatory, since at bottom *maître et possesseur de la nature* already entails *maître et possesseur de la société.*

We thus see that the basic assumptions or ideals indicated above are in wide circulation and constitute the formulas which the machine has generated out of the inherited notions of the omnipotence of reason and the attainability of earthly happiness. Those basic Enlightenment ideas were extremely complex and capable of development, pregnant with diverse experiences, passions, aspirations, and hopes. They bore fruit under the most varied practical, social, aesthetic, and logical conditions, and established amongst themselves connections having surprising, fertile consequences. More than any other epoch since that of the Greeks, that of the Enlightenment had an aura of genius about it; it was full of contentment and self-assurance, inexhaustively inventive and imaginative. Once more on Europe's fruitful soil old and new buds fertilized one another, and a vegetation of bewildering variety blossomed. As Sorel wrote, "it required a culture mixed of very powerful, distinctive, and diverse elements, for man to attain art, philosophy and religion; i.e., what today we call freedom."[19]

What all this has led to in our own time has been well characterized by Bergson. "We have seen the race for comfort proceeding faster and faster on a track along which are surging ever denser crowds. Today it is a stampede."[20] One aspect of this is something which Max Scheler imputed to all dominant social strata in history, but which is no longer exclusive to such groups: "pleonexia," signifying simultaneously greed, arrogance, and ambition for power.[21] This is a most useful term for the social-psychological characterization of our age. One might employ it to define "mass," since the connotation of "primitivism" previously attached to the term no longer seems useful. No matter what learning or what social position he possesses, an individual is part of the mass insofar as he exhibits pleonexia. Conversely whoever exhibits self-discipline, self-control, whoever possesses detachment and a view of how to transcend himself, belongs to the elite.

Toynbee has seen self-control, as well as abandon, as typical symptoms of the decline of a culture; a Cato belongs in the atmosphere of the civil wars, of caesarism, and of political gangsterism, as much as a Clodius does. Conceivably, ethical demands have their own historical location, but when their time has come they take possession of it in a particularly decisive and unconditional manner. As to what we have previously noted as our own time's astonishing absence of all ascetic ideals, Toynbee expresses himself in a sybilline manner: "We search in vain for . . . asceticism, and we may perhaps tentatively draw from this fact the cynical conclusion that, if our Western Civilization has indeed broken down, its disintegration cannot yet be very far advanced."[22]

6 Crises in Cultural Development

Problems of Mature Cultures *

WE MUST NOW bring our criticism directly to bear on our own culture, and at the same time dispose of some difficulties and objections. Our argument moves within the context set by various attempts at a philosophy of culture, dating from the first part of the century. Among such attempts, two of those most worthy of consideration were made in the shadow of one or both of the World Wars, in Germany by Spengler, in England by Toynbee. However, as P. A. Sorokin has indicated,[1] a number of the relevant considerations later developed by both those authors had already been put forward by the Russian N. J. Danilewski (1822–1885) in his book *Russia and the West.*[2]

All these authors agree on the following points. The earth has seen many phenomena, largely independent of one another, called cultures. Cultures—rather than peoples or states—ought to be considered the appropriate units of discourse in universal history. In spite of their several peculiarities, all cultures exhibit some uniformities or structural similarities. Our own culture (in essence, Western European) is now in decline, having entered the "civilizational stage," and as a "mature culture" is slowly approaching its end. This allows for significant comparisons

* *Spätkulturprobleme: Spät,* literally "late," is here translated as "mature"; it is meant to suggest a state of *advanced* maturity verging on decay. The literal translation is sometimes used in this meaning, as in the expression "late capitalism," frequent in Marxist literature.

with the later conditions of previous cultures, and particularly that of classical antiquity.

Most of the arguments and the evidence put forward to support these gloomy conclusions are so striking that they cannot fail to impress the majority of historically informed readers who are willing to take qualitative evidence into account. Of course the ability to discuss these theories presupposes a level of learning shared by only relatively narrow circles of readers.

There is a remarkable degree of agreement, among the authors who deserve consideration, about the indices which justify the use of a term such as "mature culture." In characterizing the intellectual features of a culture in its civilizational phase, one can use the expressions employed by Goethe at the end of his few but significant statements on the theme "Epochs of the spirit" (1817):[3]

> Nobody ponders wisely or operates quietly; they randomly scatter seeds and weeds. There is no longer a central point on which to set one's sights. Everyone puts himself forward as a teacher and a leader, and treats whatever foolishness he has to offer as a perfect accomplishment. . . . Qualities which by nature used to develop in separation are now brought into strident conflict with one another.

Toynbee, in the concluding part of the condensed version of his work *A Study of History* (vol. 1), lists a large number of features associated with the mature phases of cultures. Particularly significant are the symptoms of a spiritual nature, which are often dealt with through polarities; for instance, both "truancy" and "going adrift" find a counterpart not only in the sense of sin, but also in the sensitive consciousness or in flight from the world, which currently takes the form of a flight from society. Next to the archaists and those who romanticize the past are found the futurists; thus, in our own time, plebeian and proletarizing tendencies are part of the same landscape in which those most sensitive plants, Klee's paintings, flourish. Syncretism, the mixing of all styles, forms, and feelings, triumphs at all levels and in all settings.

To such aspects of the civilizational phase Spengler adds larger, imposing forces of an objectively social nature: the megalopolis; money; the rational intellect. Spengler recognized that intellect as the urbanized form of the mind; it "reforms the great religions of the springtime, and sets up by the side of the old religion of noble and priest, the new religion of the Tiers État: *liberal science*."[4] Spengler saw ethical socialism as the force which, since the turn of the century, has asserted itself as the carrier of a final interpretation of the world. The "political economy converted into the *imperative* mood"[5] had previously reigned unchallenged but now the heart has become thoroughly socialized by feelings of sympathy and sentimentality. These feelings provide a kind of touching accompanying music to our brutal cities, with their higher death rates. And the death rates demonstrate that although mankind can adjust behaviorally to the industrial city, it can no longer cope with this form of existence at the deeper, biological level of its own constitution.

Some Objections

One cannot exclude the ideas just discussed from a social-psychological inquiry simply because they have gained in the minds of many the status of something self-evident. But we must briefly consider the question of the extent to which this book is to be seen against the background of that interpretation.

Sometimes the question is not so much the truth of statements of the kind we have recalled, as their claim to being all of the truth. Such reservations are especially plausible when applied to writings particularly committed to a sustained critical evaluation of the culture of our own times. It can be argued that the systematic emphasis placed on pathological phenomena in such writings leads to a biased account. Yet such an emphasis is unavoidable. All living entities become accessible to analysis only when they are being taken apart, decomposed; and the reason is possibly that only at that point is the course of their development wholly at an end, or that at that point causal relations become simplified. The physiologist must interfere with the organism in order to experiment with it and gain knowledge of it; only in an unnatural state is his object open to inquiry. Can it be otherwise with the physiologist of culture? In the study of culture, the richest of all living phenomena, where bloom and decay continuously contend and alternately gain the upper hand, does not the student run the risk of missing those covert processes which maintain health, in his concern with the pressing problem of decay? More generally, one may wonder whether those phenomena which call for greater attention because of their urgency and their apparent significance are truly the more representative ones. Particularly in the social realm, those aspects claiming urgent attention and those most accessible to inquiry should not be taken to stand for the totality or even the majority of aspects, nor as pointing up decisive trends. In view of this, one must seek to establish the true significance of those symptoms which Spengler and Toynbee discuss, and which this book also occasionally mentions.

We admit that such an objection can be made to our

own argument, but we consider it to be only relatively, not decisively, significant. Besides, the German situation requires some additional reflection. Here, to the theory of culture, in association with sociology and social psychology, falls the task of resuming that "great dispute" which on the whole both philosophy and literature have abandoned. In a significant essay, Erich Franzen points out that modern fiction—particularly that cultivated outside Germany, and represented by such as Hemingway or Camus, is dominated by a profound feeling of anxiety. These authors think through to its conclusion, with ruthless consequentiality, the fate of a subject who finds himself utterly on his own in an uncontrollable and dangerous world: "an objective tension now obtains solely between the fragments of the ego and the lost absolute which alone could hold those fragments together."[6] This transformation of reality into a space with barely imaginable dimensions ill accords with the prospect of a rationally guided society, which must consider this kind of literature "a deplorable venture." Indeed, that society is bound to find fault with books that focus upon "the marginal areas of existence," and in which a kind of mythology or "paramythology," written as if in hieroglyphics,[7] starkly challenges the masses' belief in progress or the simple Utopias of the prophets of happiness. Exactly this kind of confrontation is particularly difficult in Germany; Kafka initiated it, but nobody carried it on after him. As Franzen has it, the great shock brought about a kind of psychological crippling; the emotional elaboration of the anxiety complex remains truncated because we prefer to block off anything that might activate that complex. The critics reject all negation of eudaemonism as "nihilism," and no intellectual light is thrown upon the situation.

Thus Franzen insightfully accounts for the tendency, in German contemporary fiction, to deemphasize the elements of danger in the situation and to insist upon romantic themes, a tendency clearly in contrast with the disquiet perceivable in poetry and in painting. Yet poetry and painting are not capable of sustained, comprehensive analysis; they tend to the level of minor art (particularly painting, with its tendency to turn into graphic art). For greater impact and reflection, one must rely upon the interpretive contribution to be made by fiction and by the scholarly and scientific disciplines, philosophy above all. But when these, in turn, fail to fulfill that function, it falls exclusively upon the science of culture to throw intellectual light upon the situation. We owe the not inconsiderable dimensions of German literature in the theory and criticism of culture to the objective mutual relations of the various cultural fields, as well as to the particular condition of the German mind.[8]

A New Cultural Threshold

When James McNeill Whistler objects to the view that ours is a time of decline, one feels like assenting.[9] After all, what some may perceive as decay, as disintegration, may be experienced instead as a liberation, the opening up of new horizons, the flight from a cage. What, from the outside, appears as a predetermined, overdue conclusion, as a progression along a preordained downward curve, may have been perceived during its occurrence as a new venture open to forward-looking forces. "Destiny is always young," says Spengler.[10]

So far, in common with many current efforts at critical reflection, we have freely drawn upon many different sets of working tools. But one can now reconsider the contemporary situation from a wider and more valid, more systematic viewpoint, and place into a larger framework the question of how "mature" our own time is.

To do this we must recall our argument that, with industrial culture, mankind has encountered an "absolute cultural threshold," initiated a qualitatively unique sequence of events, which advances at an unprecedented pace. Human history knows very few major developments whose progression is irreversible; which, as it were, place the whole story of mankind on a new level, and subordinate to their commanding influence all other laws of history. Already in the first edition of this book (1949) we argued—as we have above—that the history of culture knows only two such decisive breaks: the prehistoric transition from a hunting culture to settlement and agriculture (the Neolithic revolution), and the modern transition to industrialism. Later, in our *Urmensch und Spätkultur,* we stated:

> The impression is that the transition to industrial culture, the mastery acquired over the inorganic realm and particularly nuclear energy, open a new chapter in the history of humanity. We have been involved in this process for the last two hundred years, and this "cultural threshold" has a significance comparable to that of the Neolithic. No sphere of culture and no nerve of man will remain unaffected by this transformation, which is to last for centuries to come. But it is impossible to predict what will be consumed in this fire, what will come out of it changed, finally what will endure it relatively unchanged.[11]

A. Varagnac has recently spoken of a change in mankind's natural environment, entailing different "technical levels."[12] The hunting-and-gathering culture was associated with the animal realm; the rural and pastoral one with the vegetal realm, and finally, modern culture turns to nonliving matter.[13] In this framework one may argue that the whole historical epoch designated as the "era of culture" is about to be terminated, if one uses the word "culture" in a meaning applicable to mankind's higher cultures so far. Alfred Weber has argued exactly this:

> A process of ideal decomposition which puts everything into question is now affecting the whole earth. This ending cannot be compared with that of any other era. What is at an end is not just the development of the West hitherto; nor just, more widely, the sequence initiated when a new, conquering humanity, rode into the world around the middle of the second millennium B.C. What is ending is the whole pattern of higher cultures built up since 3500 B.C., and with it those primitive and semiprimitive cultures which had existed side by side with the higher ones.[14]

If such views have any validity, their implications are tremendous. In this perspective, the whole theme of the maturity and decline of our times becomes as it were "sublated,"* brought forward in a relative form. It becomes possible to treat the present time as embodying the interference pattern or interpenetration between a civilizational era in the old style and an utterly novel era. In the remainder of this chapter we intend to explore those implications.

One such implication is the possibility of accounting for the often voiced feeling that ours is a "time of transition." On this point, the Spengler-Toynbee theory of culture cycles is inadequate. It implies that historically untried peoples must be waiting outside the gates of the

**Aufgehoben.* The expression *Aufheben,* loaded with dialectical meaning especially in its use by Hegel and Marx, means both "to abolish," "to preserve," and "to lift to a higher plane."

aging West European–American culture—but there are no such peoples in sight. Rather, we are seeing the re-entry upon the historical stage of such people as the Indians, the Egyptians, and even the Chinese: people who, according to Danilewski, had reached their posthistorical stage, and who Spengler considers "fellahin peoples," products of the spent soil of worn-out cultures. On the other hand, if what defines a people as untried is the fact of being so much in motion as to appear formless to the casual observer,* then the whole world of today is again full of such peoples.

Indeterminacy as a Characteristic of the Times

We turn now from the world-historical problems discussed above to our social-psychological themes, having answered the question posed at the beginning of this chapter about the place occupied in our concerns by the critique of our culture. However subjectively justified by many phenomena, a merely critical posture must make room for more soundly established knowledge, possibly along the lines of the working hypothesis just advanced. If what we are dealing with is an interference between, on the one hand, a condition of civilization already seen several times on the world's stage, and on the other a unique process whereby mankind is establishing the "natural environment" in which it is to live in the future—in sum, the multidimensional process of "industrialization" with the associated changes in consciousness—then it becomes possible to place into a new framework symptoms and phenomena now to be mentioned. Within this context they

* *Urvölker durch und durch bewegt sind bis zu dem Grade, dass sie dem flüchtigen Betrachter gänzlich formlos erscheinen.*

all appear two-sided: they face both forward and backward. This two-sidedness and ambiguity is part of their very nature, and constitutes a distinctive feature of our times.

One often feels that current states and events result from wholly heterogeneous components; that there is something ambiguous and *objectively blurred* about them. Even very significant phenomena, as Thomas Mann well knew, oscillate between the serious and the farcical; and technical necessities, with *their* associated mentalities, can be anything from purpose-grown, historically developed constructs to realities endowed with legitimacy, deeply rooted in the heart. Only a few modern painters have been able to represent this state of reality, by producing surrealistic images which are *objectively indeterminate.*

Is it war we are having, or is it peace? Do we have a fatherland or do we not? Do we live in the era of socialism, or of capitalism? The answers are arbitrary, but not because they vary with one's viewpoint—rather, because both answers are equally correct. Consider a country where the classical revolutionary postulates of the Gotha program have been 80 to 100 percent fulfilled,* where the living standards of the lower classes improve all the time, *and* where the rich get richer all the time, too. What are you going to call such a country? How do you characterize a time when the socialization of feelings has progressed to an unprecedented extent, so that everybody agrees that all sufferings, real or imagined, should be succored, and when, on the other hand, the distances between classes grow objectively greater?

Other examples could be drawn from numerous

* The "Gotha program" (1875) was the ideological charter of the German social democratic party in the latter part of the nineteenth century.

spheres. Some contemporary students of internal medicine unreservedly believe in the psychic causation of organic illness, to the point where illness appears as guilt and treatment nearly changes into metaphysics. If one does not go so far, then one remains tied to a reality which is itself objectively blurred, objectively indeterminate, and which shares with the previous sociological examples the property of allowing for conflicting judgments. Slowly such realities are turning into a distinctive feature of our times; consider that even physics now envisages entities that are blurred, unclearly related, and objectively indeterminate. What of art, then? Here we encounter, as for instance in Klee, images that decisively violate the boundaries art previously imposed upon itself. Like exotic plants, they astonish us with unprecedented mixtures of qualities, surprise the eye and the sense of touch. Sinuous, sintered, shimmering, chalky sedimentations suggest the existence of new chemical elements. Were its public endowed with a different, more archaic consciousness, such art could well be perceived as a direct rival to nature, as a fetish.

The consciousness of our times pursues with extraordinary energy its preassigned task to violate boundaries, to commingle heterogeneous elements, until it assumes these features itself. The previously discussed "opinions" (see, in chapter 3, "Opinion as Second-Hand Experience") again become relevant in this context. Undoubtedly they are formed by a compulsive reaction to the loss of direct experience and to the unpleasant tensions of ignorance; yet they also represent a way in which objective indeterminacy reaches into the individual mind, translates itself into everyone's thinking. Clearly the need for precise information and sharp thinking has become less widely felt, and is now limited to narrow circles. It is remarkable how well the widely diffused need to entertain and express opinions fits with the actual indeterminacy and vagueness of reality itself.

A further example shows how certain social-psychological facts, once looked at in this framework, appear two-faced, almost ambivalent, although they are given only one simple, fatalistic significance in Spengler's theory of culture. Spengler, as is well known, attributes a marked *rationalism* to mature epochs. He thus sharpens a polemic originating with naturalistic-Jacobin tendencies (Rousseau and Diderot), then carried on by romanticism and pietism, and later clinched, as far as German public opinion is concerned, by Nietzsche.[15] It is undoubtedly the case that every culture has seen a long-term trend "from myth to logos"; yet one can definitely assume that the future *no longer* holds any prospect of a resurgence of mythical consciousness, since the industrial culture presently conquering the globe is rationalistic through and through. The question of the origins and consequences of the rationalization of the human spirit is undoubtedly the key one. Max Weber had already posed it,[16] and we intend to contribute to it a few considerations.

Keiter has recently enunciated a general trend within universal history, to the effect that, on average, preference ends up being given to more effective and less demanding means of action. But according to Keiter this trend applies also to patterns of thought, so that in the long run (and not without setbacks) more correct, complex, differentiated, effective thinking patterns find their way to the top. To quote Keiter: "This entails, in principle, that the course of human events cannot ultimately reject many of those we consider signs of a 'higher culture,' since these make their way forward on ground of 'technical progress.' "[17]

This suggestive view entails that the social-historical process possesses a kind of self-regulating mechanism which in the long run aims that process in the direction of the most effective and expeditious processing of all data—not just natural data, but also social-historical ones! The end result of this process would undeniably be the progressive "rationalization of the human spirit" in the course of universal history.

Some anthropological considerations clarify this conclusion. It is part of the essence of language that it places upon its distinctive level, that of sound, *all* events, internal as well as external: dreams as well as outside processes, feelings as well as tools. We do not reflect enough on how remarkable this is. Insofar as we articulate words to indicate both internal and external data, we are in possession of a universal instrument. This is a most powerful convenience, since one could well imagine us having available a language of sounds for external events only, and having to manifest internal ones only through a language of gestures. Instead, we are capable of a process of abstraction involving a most significant *facilitation* effect, since it allows us to disregard the *practically* most significant distinction of all, that between inside and outside.

Yet, on that single level, language allows innumerable emotional differences; it expresses through several devices, among these verse and rhyme, the varying distances of things from our hearts. Modern rationalization, the end product of a long development, entails not only that internal and external phenomena are placed on one and the same level, but also that basically they are placed *at the same distance.* Otherwise put: it becomes possible to assume, vis-à-vis *all* things and events, the same behavioral posture, a matter-of-fact one, addressing only objective characteristics. This permits an extraordinary extension of a mode of conduct which from long habituation we take on relatively easily: rational theory and practice. When we say that one can "deal with" not only cars, but also atonal music or a factory's social climate, this has a cynical sound. We intentionally arouse this impression to indicate that the long-run task of humanity is to determine which realms of things are, and which are not, amenable to rationalization. Only trial-and-error can give a response, and thus determine also what is to count as cynical. It is not possible to make exact predictions. For instance we have now attained, in the sphere of the state, a degree of rationalism and of matter-of-fact orientation that Fichte or Hegel would have found outright cynical. Equally, we have practically disposed of the time-hallowed, pathos-laden concept of art as a lofty, autonomous realm of reality—a concept which was the norm of aristocratic art. In other spheres, too, the key consideration becomes which "differences of distance from the heart" must be preserved, if dehumanizing consequences are to be avoided.

Such reflections clarify the above notion of a self-regulating social-historical process aiming at the most efficient and convenient, or at any rate optimal, *processing of its own data.* The advance in question will be protracted, checkered, risk-laden, and perhaps bloody. One day, perhaps, the critique of culture will assist in that advance, but for the time being it largely seems to content itself with emotional protest on behalf of tradition against what are perceived as excesses of the rationalization process. Yet the arguments in this chapter already show that simply to characterize the present times as "mature" does not throw adequate light upon all contemporary problems; in particular, such a view has little to contribute to the above discussion of the problem of rationalism. Perhaps the whole theme of "culture," particularly in the German sense of the term, belongs in a certain sense to the past. Indeed, not just the theme, but rather its very referent—the period of preindustrial high culture with all its associated rhythms of existence—appears now as a unique event, lasting from 3000 B.C. to 1800 A.D. In this perspective it no longer appears indisputable that culture in *that* sense embraces the highest values and experiences of meaning.

7 Social Psychology and Psychoanalysis

Some Remarks on Freud's Undertaking

SOME READERS MAY have wondered why our "phenomenological" descriptions of social-psychological phenomena have made no use of models from depth psychology, or of the concept of the "unconscious." In fact, we feel that psychoanalysis is currently past the peak of its influence—at any rate in central Europe—and survives not so much as a theoretical construct but as a perspective. Furthermore, while psychoanalysis does indeed have implicit social-psychological premises, these are not enough to strengthen any claim that it be used as a social-psychological doctrine. All the same, psychoanalysis is undoubtedly significant enough to warrant some discussion at this stage.[1]

It is to be expected that a psychological theory will be shaped by the empirical material available to it within the sociological and social-psychological context wherein it develops. For this very reason, in this century psychological research has developed mostly on the basis of very biased materials. For its part, the classical experimental psychology of the past century reveals in many of its features its origins as the unconscious self-analysis of the academic estate: in particular, in its emphasis upon intellectual functions as a subject of study, in its conception of sensation as a *cognitive* element, in its tendency to intellectualize emotion and volition, in its denial of the subconscious, and finally in the dominance of a view which derived the conviction of the existence of other persons only from complicated arguments.

On the other hand, when Freud was able to point to boundless "pleonexia" and uncontrollable desire as the background to his patients' conflicts and difficulties, this showed to what extent those patients mirrored the general conditions of the culture within which Freud himself operated. His painstaking, imaginative, clear-eyed clinical observations synthesized and generalized those conditions into the "pleasure principle": "It seems as though our total mental activity is directed toward achieving pleasure and avoiding unpleasure—that it is automatically regulated by the pleasure principle";[2] or again: "This principle has unrestricted sway over the processes in the id."[3]

These conclusions, together with those drawn from the analysis of dreams, where instincts appeared to determine men's consciousness, thought, and conduct, constituted the foundations of Freud's psychology. Contemporary conditions were reflected in the individual's pleonexia, drivenness, and impersonality. Psychoanalytic theory is like a high-resolution lens, which most effectively focuses on properties such as the above ones but is unable to determine their origins.

As far as these general views are concerned, there is really not such a great difference between Freud and some other psychologists, Nietzsche for instance, who operated with the same material—members of the upper-middle class in large cities around the turn of the century—and attained results comparable to Freud's, sometimes even in their details. Freud's unique standing is due, beyond that, to his outstanding productivity as an empirical researcher. He was the first to discover and traverse wholly new inner landscapes, which he charted and described, thus putting all later travelers in his debt. He spectacularly widened our sights and strengthened our capacity for judging human nature, by mapping the ways and (above all) the by-ways of the soul. One can only agree with William McDougall both in his rejection of almost all of Freud's dogmas and in his overall judgment of him: "Freud has, quite unquestionably, done more for the advancement of our understanding of humanity than any other man since Aristotle."[4]

Freud himself showed remarkable caution when it came to interpreting higher mental activities, and art in particular, on the basis of models developed to interpret dreams and neuroses. By and large he contented himself with emphasizing a few "motifs," derived by means of his analytic method, and evidenced in classical works of art. It is only since his time that psychology has learned to rival, in insight, depth, sophisticated description, and interpretation, the knowledge of man demonstrated by literature. Thanks to Freud, the huge gap which, as late as the 1890s, separated academic psychology from Chekhov, for instance, was considerably narrowed. Consider, for example, Freud's presentation of the demonic trait in his discussion of personality patterns in *Beyond the Pleasure Principle:*

> We have come across people all of whose human relationships have the same outcome: such as the benefactor who is abandoned in anger after a time by each of his *protégés*, however much they may otherwise differ from one another, and who thus seems doomed to taste all the bitterness of ingratitude; or the man whose friendships all end in betrayal by his friend; or the man who time after time in the course of his life raises someone else into a position of great private or public authority and then, after a certain interval, himself upsets that authority or replaces him by a new one; or, again, the lover each of whose love affairs with a woman passes through the same phases and reaches the same conclusion.[5]

Some people have been led to considerations of this kind by their own experience, others by fictional or dramatic literature; but before Freud they had never been put forward on the basis of methodical, systematic research, even if one admits that Nietzsche's work was even richer in scattered individual insights.

Althought, as I have said, one is impressed by the master's own caution, Freud's students showed themselves all too eager to bring his methods to bear upon the largest problems, and laid their traps for the biggest game. Otto Rank undertook to account for the whole process of artistic creation in psychoanalytic terms. As he himself characterizes his own views: "Psychologically speaking, the artist lies somewhere between the dreamer and the neurotic."[6] "Such individuals, the artists, become aware of conflicts which cannot emerge into the consciousness of normal subjects, and project them into their own Ego . . . *at a point where such conflict is too advanced for dreams, but has not yet become pathogenic.*"[7]

On such theoretical bases it was contended that there is no substantial difference between, on the one hand, the dreams and dream-books of simple souls, or the scribblings of neurotic literati, and on the other hand those immortal works of the masters which embody the spirit of whole nations. Only within a certain social system is it possible for such an opinion to be even conceived; and in fact it has nothing of significance to tell us as far as psychology is concerned, but is very revealing as to its sociological background. The man who labors at piece-work in the day, goes on laboring with dreams in the night (to compensate for unfulfilled wishes)—only the "laborer of the mind" keeps at it in the daytime as well.

Naturally such theories cannot carry much conviction.

They may make this or that point correctly; but to treat such points as the whole, to equate children's drawings or dream stories with art (or vice-versa) on the basis of a few similarities, amounts essentially to "a paradoxical and thus unexpected subsumption of an object under an otherwise heterogeneous concept"—which happens to be Schopenhauer's definition of the ridiculous.[8] Furthermore, fruitful and imaginative as dream theory may be, men shouldn't take dreams utterly seriously. Already Calderon stated that life itself was but a dream, but was unwilling, on this account, to attribute validity to dreams: "and the dreams themselves are but a dream."

By and large, the numerous applications of psychoanalysis to the history of ideas and to art have not rendered a good service to those disciplines. Cultural values possessing the emotional charge and the traditional significance of the Goethe complex (however impoverished and conventionalized) were cheapened, almost turned into farce, by being treated in a spirit of paradox and low humor. One could detect a breath of that "hurricane of farcicality," which according to Ortega y Gasset has made itself felt, in Europe, "everywhere and in every form."[9] On the contrary, Freud himself showed, sometimes by just a few words, how desperately in earnest he was, how aware of the contingency and frailty of knowledge, and how willing to accept corrections to his views. He admitted that even a process as central as displacement is not understood in detail; he spoke of "the obscurity that reigns at present in the theory of instinct,"[10] and he even called that theory of instincts, the core of all his labors, "as it were, our mythology."[11] Already in *Beyond the Pleasure Principle* he claimed no specific psychological import for the hypotheses concerning instincts, but rather the significance of a "metaphysical speculation."[12] His theoretical cause would have been served best by seeking to support empirically his later hypotheses concerning the conservative, regressive, and stereotyping nature of certain classes of instincts, in particular the so-called death instinct. However, once it was seen—and correctly so—that some of Freud's discoveries and concepts were not merely aids to therapy but could be more generally applied, some of the ensuing applications were too farfetched and too facile.

Social-Psychological Uses of Freudian Theory

In the half-finished state in which the master left it, Freudian theory raises a few major problems. Hofstätter rightly remarks that the Freudian legacy has made a greater impact upon social science than has any other psychological doctrine.[13] The principle of ambivalence, disclosing the two-sidedness of many emotional bonds; the doctrine of the superego, and that of authority relations—these are only some of the features in Freud's theory which have made such an impact. Freud was the first to emphasize the long-term significance of socialization tactics, of indulgence or severity in dealing with children; above all, he was the first to establish the methodological legitimacy of intuitive interpretations. However, such successes are mostly a matter of applying concepts from individual psychology to social-psychological problems; whereas all those Freud expressly intended as his social-psychological contributions must be counted as failures. The theory of the primitive horde contained in *Totem and Taboo* fully deserves McDougall's derision; and all later contributions of Freud's are untenable in their core contention—the directly or indirectly sexual nature of all social relations. According to this central position, all social activity is either expressly of a sexual nature or is "the expression of aim-inhibited libido, i.e., of energy derived from the sex instinct but diverted into channels of expression other than the explicitly sexual."[14] In Freud such absurdities appear plausible because they amount to a perspective from which his readers may gain insights into several significant points; but all too often those readers forget how arbitrary and fragile that perspective is.

Here and there one finds passages in Freud's writings suggesting his own awareness of such weaknesses. At one point, for instance, he mentions quite briefly the "social emotions . . . determined by showing consideration for another person," but unfortunately without going into "the origins of these social impulses and their relations to other basic human instincts."[15] This is regrettable, especially since it leaves unclarified one of his central topics, the theory of neurosis, where neurosis is characterized as involving "the preponderance of the sexual over the social instinctual elements"[16] and described as a kind of illness of the "social organ." This sounds suggestive, and one would like to hear more about it, particularly because elsewhere neurosis is defined otherwise; for instance, as a conflict between ego instincts and sexual instincts.[17]

A further difficulty is associated with the previous one. In the seven-page-long index appended to the famous *Introductory Lectures* there is no entry for "will." This again indicates how one-sided was the material selected for attention. Not only was it derived, understandably, from drive-obsessed psychopaths, but it also emphasized socially isolated and institutionally maladjusted subjects. Yet the higher, culturally significant—metapersonal, as it were—forms of the will are the necessary products of general conditions, which surpass and surmount the individual. In persons placed outside the reach of social demands and imperatives—and those who have become lawless—it is no longer possible to detect the boundaries of human plasticity; and it is not surprising that where the automatic mechanisms concerned with pleasure and unpleasure operate without restraint, there is no trace of the will.

The thematic gaps characterizing psychoanalytic research and theory, and originating in the narrow historical and social location of the emergence of the doctrine, are evident from a further observation. We have remarked on the caution shown by Freud in approaching certain topics; this caution is not in evidence as far as religion is concerned. *Totem and Taboo,* intended to discuss the early stages of religion, is probably the weakest of Freud's writings, since it seeks to detect in all archaic forms of association the patriarchal nuclear family, from which Freud derived the identity he constructs among God, King, and Father. Freud could understand only the monotheistic patriarchal style of religion, and for this he felt a strong aversion, which made his polemic unfair and malicious.[18] Thus, in his struggle against "God the father," he saw a nostalgia for father as the root of all religion; he derived the contents of religious experience from the wants and needs of childhood prolonging their effects into maturity, and treated them as an attempt to master the sensible world in which we live by means of a world made up of wishes.

Now, all this may be plausible when it refers to the subjective motivations of some infantile individuals, but it misses a number of plausible and relevant questions which the psychologist could pose concerning religion in general, not just Judaeo-Christian religion. At any rate there is no case for reducing religious behavior to a narrow complex of internal determinants. Since there is no one specialized religious "organ,"[19] and since religious motivation can enter into almost any form of human conduct, no matter how oriented, religion can fuse into unity the most different actions of the same individual or of different individuals. To that extent religion belongs, with language and labor, among the great "integrative" forces in human existence, those forces which bring together and unify. It is hard to see how such a function could be served by forces as intrinsically unruly, contingent, and divisive as the wishful imaginings of individuals.

Furthermore, religion constitutes an answer or reaction to certain specific conditions of a spiritual nature, and as such can never be attributed exclusively to the operation of subjective feelings or drives. Consider the following "primal phenomena" (*Urphänomene*): that the world constitutes both a transitory fantasy as well as the final, ultimate fact of any given moment; or that each consciousness experiences a contrast between a "world" accessible to everybody and its own solitude, a contrast due to man's lacking that tight fit with the environment, characteristic of animals. Only religion can come to grips with these puzzling ontological facts in terms other than the purely intellectual. Science can hardly throw much light upon them, yet it is difficult to convince people that these facts are irrelevant.

It is usually said that religions work out their answers from narrow perspectives, which they try to extend to the world as a whole. There is little doubt that this is so; yet, far from being a deplorable mistake, the extrapolation of limited experiences, and above all of limited models, is a general ontological law of development, also valid outside the human realm. Therefore, religion should possess historical substance, it should be not just the interpretation of something that happens, but an event in itself. Yet problems of this significance cannot begin to be raised in the light of the concept of wishful thinking, with its derivation from the psychology of spoiled children.

Our critique can be synthesized as follows: Psychoanalysis works well at the level of the individual, badly at the level of the society. Its findings are convincing when they concern individuals whose "social organ" is diseased, that is, neurotics. It is all the more applicable the more fragmented the society is—which means it is applicable to a considerable extent in our own society. Yet categories derived from the psychic existence of individuals become oddly deformed and ultimately fall apart when one tries to load them with contents referring to the collective level. For example: "It is only with the elevation of the never-forgotten primal father that the deity acquires the features that we still recognize in him today."[20]

The Pleasure Principle and the Principle of "Unmasking"*

The basic tenet not just of psychoanalysis, but of all "unmasking" psychology since the Greek Sophists, is the pleasure principle: the notion that in the final analysis each individual acts as he does in order to attain pleasure and avoid unpleasure.† When society has decomposed to the extent that men's acts are no longer understandable on the basis of their social functions but only on the basis of each man's concern with himself; and when as a consequence

* *Lustprinzip und Entlarvungsprinzip.*

† The awkward expression "unpleasure" has been given canonical status in the official Hogarth Press English edition of Freud's works.

mutual distrust sharpens the observation of human actions—then unavoidably the view triumphs that the ultimate meaning of all action is to be found in a single and egoistic intent.

We find Hofstätter's objection to the pleasure principle convincing: pleasure itself entails a reference to something which transcends the individual, and necessarily presupposes some commonality with other persons or with personified things.[21] In any case, the doctrine of the pleasure principle is put forward in association with other theses: in particular, with the view that values and ideals are but clever mechanisms giving expressions to the subconscious or to the life process (Nietzsche), or to some other factor of an impersonal nature—according to this view, ideal constructs are mere "ideologies" in which the above interests become clothed. Or they may be sheer illusions, hinting at processes whose positive import we fail to recognize.

Such theses are not acceptable for some deep-lying instincts of an ultimately moral and social nature, and thus encounter *resistance*. Some psychoanalytic authors have been clever enough to take into account such resistance, and to explain it in their theories as a necessary complement of the very mechanisms which are being uncovered. But the same resistance confronts all theories where *self-deception* appears as a central process on which light is to be thrown. According to such theories, some people will use technico-psychological means to manipulate others into behaviors which the others' illusions would not allow them to adopt, and which are instead in keeping with the illusions of their manipulators. This is irresponsible, but a psychology of this kind, once it has become popular, cannot motivate a resistance to such irresponsibility. It entails an image of man which is variously unacceptable, in that it cannot be lived. If someone is to take responsibility for an act, it is necessary that he should feel the act belongs to him. But he cannot feel this if he is made to think that his inner motivational commitment to the act is a delusion, behind which lies a purely objective process advancing under its own momentum: a process that increases pleasure, or any other mechanism operating purely at the service of the ego. If one really makes such a view his own, one must cease to consider oneself as capable of choice. Such theoretical schemata are equally unsuited to becoming *motivations* of our *immediate* social conduct, insofar as the latter presupposes durable and sustained relations among people who must exist together.

To say this is not to advocate well-meaning moralizing in psychology; should anyone, to that end, indulge in the usual derivations of the first class, the words of Callicles must be opposed to him: "Somehow or other, Socrates, your words always appear to me to be good words; and yet, like the rest of the world, I am not quite convinced by them."[22] However, it is worthwhile to outline the following situation, and to sketch a social-psychological interpretation of it.

A component of that vast and complex process which we summarily call the dislocation of the old social order by industrialization, was a progressive devaluation of *immediate* social contact, the significance of which had already been somewhat reduced by the conventions characterizing the estate system of stratification. The general demand for "naturalness" of demeanor is merely the other side of the same coin. Interpersonal relations become objectively, sociologically, and thus also psychically, *reified.* States of feeling such as "magnanimity," "piety," "loyalty," cease to be expected of people, and a rational orientation of persons to one another comes into its own. It is as if the whole society were taking on that experimental, detached attitude to others which only a few lone thinkers had previously projected. Once the Other begins to be rationally contemplated as a "case," the process of increasing rationalization of patterns of thought sets in. Ultimately we shall describe the Other's behavior by means of value-free concepts of causality and finality, abandon as irrational the very concepts of person or indeed of action, and treat behavior as activated motion or conditioned reflex. Accordingly, if people are found making reference to meanings and motives, this phenomenon can be treated as having purely instrumental significance: it intervenes simply to facilitate adaptation, coordination, social integration. Even the pleasure principle can be treated as such an intervening mechanism.

A psychology of this kind can come to full fruition only on the assumption that immediate social solidarities and the ideal models developed within such contexts, have been destructured or thoroughly restricted in their significance. For, as long as such models remain in operation, they protect their objects from the advance of analytical rationality.

After two generations of disappointed expectations, several circles seem to have turned away from psychology;[23] to some extent this is due to the fact that, as we suggested, some arguments and theories put forward by psychology *cannot function as motives,* or at any rate as ethically acceptable motives. For "unmasking" psychology seeks ethical legitimation in the postulate of "freedom from prejudice," which lays bare the incompatibility between ultimate value positions. But on these terms, how can one argue with someone who says: "Why should one not declare oneself for prejudice against those who claim to have none? Why should one be more of a free-thinker in one's head, than one's heart can tolerate in the long run?"

It is equally significant that one cannot subscribe sincerely to that imagery of mechanisms and complexes, and go on acting as a responsible individual. It is not possible for one to retain one's integrity as a person while seeing oneself caught in a game of self-deception—a game where what "really" goes on is very different from what appears, and even that appearance is partially a product of self-deception. Finally, it is not possible to form a framework for immediate social conduct from such ideas, to design an image of the social order, or project specific social arrangements. Anyone *not* seeking to observe, heal, or manipulate others, but seeking somehow to make others share in his cause, can find no guidance in those ideas we have discussed.

C. F. von Weizsäcker raises the following question:

> How do I relate to a fellow human being, when I treat him, in thought or in action, as a mere object? . . . So far as I can see, the personality of my fellow man imposes no theoretical restraint on the application to him of the principle of causality or on the carrying out of experiments. Whoever ceases to address his fellow man as "thou" deprives himself of the decisive experience his presence offers. Yet nothing prevents me from gaining such experience of him, and then subsuming that experience, as well as that gained purely through descriptions, into an exclusively causal schema.[24]

It is no longer possible to renounce the objectifications that have become possible; experimental psychology, in particular, cannot cease considering the Other "in thought . . . as a mere object." But this objectification, also unavoidable in medicine, becomes dangerous as it becomes adopted in wider and wider circles. One can only hope that the popularity of psychoanalysis will be the last evidence of such an attempt to gain wide circulation for scientific views, the last to extend its reach beyond the sphere of learning and specialized knowledge into the realm of publicly entertained world views. We think that the advance of experimental psychology depends on the increasing employment of mathematical methods. This on the one hand would be in accord with the trend toward abstraction described in chapter 2; on the other, it would restrict the communication of research findings to smaller, mathematically knowledgeable, circles. Since scientific psychology has not as yet much profited from its urge to interact with other disciplines, nor made much of an impact upon education or even politics, this new development may well, on the contrary, substantially favor its development.

8 Automatisms

Schematization of Behavior

IN THE PAST, many psychologists have argued over whether or not the operations of the psyche proceed according to mechanical laws, and various attempts have been made to construct mechanical models encompassing psychical phenomena. It is strange that in this context little attention has been paid to that widely diffused form of mechanization that consists of the formation of routines, of behavior patterns automatically activated by certain stimuli. The phenomenon of routinization is seen everywhere, since it is connected with the progress of the division of labor; yet its bearing upon the wider question of the mechanical nature of behavior has been seen only by Bergson. Even Freud, in his thorough and masterly investigations of the psychological processes evidenced in the wants and dreams of the neurotic inhabitants of great cities, failed to pay attention to this theme.

In our social capacities we often act "schematically," that is, we enact habitualized, well-worn behavior patterns which unfold "by themselves." This can be said not only of behavior of a practical, external nature, but also—and primarily—of the internal components of behavior. The formation of thoughts and judgments, the emergence of evaluative emotions and decisions—all these things are largely automatized. For this very reason, often they cannot be meaningfully referred to the person seen in his individuality, but rather to his capacity as role holder in a given context, in which capacity each individual person is relatively interchangeable with others. The fact that, for instance, someone is a garage mechanic means essentially that his habit (meaning both his duty and his inclination) leads him, one day after another, to the same place of work, to the same rather specialized activities. Society has an interest in his carrying out those activities, and considers primarily not what is individual to him, but what goes with the fact that he is a garage mechanic. He may well consider his work as a source of honor to himself, but this honor is itself a function of the standing enjoyed in society by mechanical arts, and is reflected upon him to the extent that he has mastered those arts.

We have thus described a phenomenon which we may call *reification* (*Versachlichung*); it consists in the fact that this individual draws the standards of his self-respect not from his own utterly individual particularities but from the objective nature of the activity to which he is committed. Again, the standards whereby that activity is judged "good," or otherwise, do not lie in him, but in the laws of the matter at hand, and the related social claims and expectations. His duty is determined by the wishes and demands of his employer *and* by the muted and stubborn commands of the material objects he deals with. And if it is true that his inclinations and his sense of duty lead him to work, these considerations reflect in turn the expectations and rights toward him vested in others.

Such socially oriented automatisms perform very important facilitation functions. They make habitual the conscious operations, including attention, involved in work, and under these conditions that work becomes equally habitual and largely ceases to induce quick fatigue. In the normal course of this operation no decisional effort is required, no affects have to be inhibited, no conflicts manifest themselves, no interferences occur between aspects of the operation. Once habits have become so specialized, the stimulus threshold becomes lower and lower, the capacity for visual and tactile discrimination is increased, motor reactions become subtler, the capacity for judgment more sophisticated; in sum, acquired capabilities increase. We can, with Veblen, call these capabilities "immaterial equipment," and consider them as a necessary product of society, as the immaterial precipitate of society's present and past experience, possessing no existence apart from the life of society.[1]

The importance of automatisms for facilitation is also obvious in the intellectual sphere. The whole realm of intellectual work—its basic features, laws, factual regularities, and individual applications—becomes habitual. With time, less and less learning effort and expenditure of energy are required of the individual, these being replaced by mental associations and by patterns of thought attuned to low stimulus thresholds and to fine discriminations, which in spite of this operate schematically. These reified and automatized operations of thought resist criticism and are immune to objections; they share these properties with all habituations, even such low-level ones as those of motor functions, where one sees a resistance to attempts to take apart and rearrange their individual component movements. Such invariance, when applied to intellectual and emotional habits, is the condition of all reliable tradition and self-reproduction, and thus constitutes a social cement of the greatest significance.

As is well known, all automatisms (except the compulsive ones of a neurotic nature) become reliable by placing themselves outside the reach of consciousness and "sinking into the subconscious." This applies by analogy to the automatisms of consciousness themselves, that is, to practical, theoretical, moral "situational formulas" (*Situationsformel*). These operate consciously yet unthinkingly, that is, subconsciously from the standpoint of an even higher form of consciousness on which they depend; they, too, resist criticism, particularly when they are socially supported and relate to behavior which bears upon society's needs. The individual we consider "superior" is one who, confronted by cognitively and morally complex situations, is capable of determing *which* require the suspension of schematized handling and need to be attended to with an eye to novel or exceptional solutions.

Even moral problems involving complex intellectual considerations must be handled in an automatic or semiautomatic manner if they are to be confronted without perturbed reflection, without loss of inner security and self-respect. An insecure moral posture is characterized primarily by inputs of reflection and by shifting motivations, followed by the shunting of conduct down to a lower level of automatisms. Correspondingly, what was said previously about automatization as a condition for lowering threshold values, for refining discrimination, etc., can be transferred to the ethical level. Here the uppermost layer of social behavior appears characterized by a higher automatism: preferences are exercised without specific choices, as if in keeping with previous decisions, without engaging reflection and abstract control, yet also while responding to the smallest nuances of a given situation.

Let us visualize a society consisting of a plurality of coordinated sets of habitualized attitudes of this kind, each controlling behavior all the way from opinions and motives to outer conduct, each applied to a distinctive institutional sphere and a different context of association. What we then find at the intersection of several of these social coordinates is a thoroughly reified and depersonalized, "functioning" individual. This individual confronts each of the various spheres of civilization and society in a distinctive routinized, durable, consistent posture, which ideally is always the "correct" one, that is, one which minimizes social and objective friction. In such a rationalistically organized society, the individual has the greatest chance of receiving general social approval when he develops a line of conduct which is at the same time objectively suitable and conventional, that is neither novel nor personal. This is the most significant quality which habitual attitudes may have: thanks to it, they will fit the requirements of the various social systems of reference and be in accord with the given configuration of the social order and of the division of labor.

Within such a system of reliably functioning mechanisms there is little place for what is distinctive about a given person, whereas the individual appears optimally suited to the multiple and diverse aspects of the social system. There is little doubt that today's highly rationalized and thoroughly bureaucratized society expects the person to develop, to a large extent, into a "functionary." Personal characteristics which hinder such a development appear unwanted, no matter whether possessed by a genius or by a socially maladapted individual. Wherever the basically unstable society of the industrial era attends to the primary socialization of the individual, its prime tendency is to his "functionarization"; and if it is possible to say that each major social order is embodied in a distinctive, representative, proverbial (as it were) human type, the "specialist" stands for ours.

Psychological Assessment

Contemporary social conditions have produced not only psychoanalysis but also another branch of psychology: psychotechnics, that is, research into and testing of individual properties, qualities, and aptitudes, insofar as they bear upon a previously defined task. The investigations in question may be very specific (for instance, when they refer to narrowly defined technical abilities), but, particularly in the United States, they often address quite generally the individual's capacity to adapt to any given social demands—a capacity which is implicitly taken as providing the "correct answer" to the problem of how to organize one's conduct.

Undoubtedly, society requires innumerable highly specialized, interdependent performances; and each individual's existence is facilitated, both intellectually and morally, to the extent that he finds himself precisely located in the social system and thereby functionally connected with its other locations. The requirements objectively flowing from the nature of the tasks at hand, and the uniform expectations society entertains vis-à-vis the role occupants, can become tied together into specific "regional moralities"—the ethos of the judge, the physician, the academic, the specialist, etc. This, however, does not make impossible a condition of total bewilderment and confusion concerning core values; and barbarism can hold sway over the whole system, even though most men act decently in terms of their own convictions, each in his place within the system.

In the psychological assessment of individuals in terms of their qualifications, however these may be ascertained, the first criterion is that the performances in question should be functionally significant for society and should entail the development of the highly efficient automatisms described previously. One seeks to ascertain not just whether the individual possesses certain aptitudes and qualifications, but also whether he is likely to attain the requisite degree of rectification. The answer to this latter question depends in turn upon the presence or absence of such traits as oversensitiveness, ruthlessness, eccentricity, tendency to criticize and dispute, lack of inner security, proneness to anxiety, hypercritical mentality, absent-mindedness, lack of openness, lack of punctuality and reliability, and tendency to challenge authority. Such additional traits as pride or audacity may be taken into account, as well as submissiveness or "pluck." Many such properties belong together, insofar as they constitute "adjustments by defense,"[2] secondary adaptations to social demands which allow a certain amount of elbow-room to egocentric tendencies.

Such properties do not arise from nothing, but rather constitute the outcome of complex and poorly understood processes—though, thanks mainly to Freud, we do know something about reaction formation, compromise, fixation, infantilism, ambivalence, etc. All those properties have in common the fact that they hinder the development of facilitating automatisms, with their attendant internal and external reified orientations. The subject's attitude becomes stably or unstably antisocial; or else, in order to prevail over such tendencies, he must both make serious demands upon his own will and emotions, and impose heavy burdens upon his associates.

Conversely, expressions such as "honesty" and "reliability" should be considered as aspects of the formation of objectively required and socially expected routines of action, of the kind discussed previously. Thus such properties are best inferred in negative terms, as the absence of the troublesome features listed above. Ludwig Klages, in his essay "Stammbegriffe der Charakterkunde," has presented a taxonomy somewhat wider than previously available ones, and has remarked that "psychodiagnosticians" are often asked whether a given person possesses honesty, trustworthiness, capacity for adaptation, etc.[3] This is understandably the most frequent and justified concern of those who pay for psychological assessments, as it bears on the question whether, in his handling of other men and of objective situations, a given person can consistently be treated as a resource.

Whether negatively or positively evaluated, all these should be considered as *consequential properties*, that is as crystallizations of "facilities" (*Anlagen*) by means of only partly understood processes. Such crystallizations occur only in the framework of determinate social, historical, and social-psychological premises and requirements. The "facilities" themselves are not directly the objects of knowledge, but must be postulated. Thus, the properties in question are genetically secondary, and appear as the outcomes of complex and protracted preconditions when viewed from the standpoint of individual psychology. In their social significance, however, they are primary, since they play a direct and considerable role within the interplay of social forces. Thus they are of the greatest interest for those who wish to remain aware of the current tendencies of public life. Klages, in the essay cited, considers properties such as honesty and adaptability *not* as characteriological features, but as consequences of "facilities," or as properties resulting from the relationship between the property-bearers [*sic*] and the community. As indicated, it is exactly in this capacity that they are of prime significance for social-functional diagnosis and prognosis; and the same applies to their opposite qualities indicated above.

Thus, this line of psychological research is far from revealing the "essential core," the specifically individual element in a person. Its typical products are statements concerning the probability of attaining a given, objective level of performance, the probability of certain claims, demands, burdens being met. If, on the other hand, one seeks to ascertain and describe the irreducible, vital component of an individual personality, one places oneself outside the boundaries of science, in a position of "unfair competition" with art. We stand by the view that a *science of the individual* is a contradiction in terms; psychology can at best aim at types, no matter how many particulars it accumulates. Every complex of properties derived from a given person, as soon as formulated, becomes applicable to many individuals. Only art is competent to portray the individual in his uniqueness; this applies in the first place to portrait painting, and probably also to literature and within certain boundaries to historical description, insofar as it focuses upon the necessarily unique deeds and products of a person in order to identify what is utterly individual to him.

Specialization and Cultivation*

As everyone knows, what gives such a thoroughly rationalized outlook to contemporary social existence is the fact that it revolves largely around sets of machines. This makes it necessary for people to engage in narrowly

* *Spezialisierung und Bildung.*

bounded, specialized, and exacting activities; for them to operate, at the points of intersection between the various machines, according to the typical ethos of functionaries, which enjoins upon them not selflessness, but depersonalization. One may characterize the attendant processes of automatization and reification by saying that human conduct itself confronts the person as an aspect of what we call the "environment," as do all related ideologies, routinized motives, convictions, and self-interpretations. Accordingly, activity which originates instead from emotions, from personal character features, temperament, etc., can only be frowned upon; the concepts referring to such determinants "cannot construe motivational structures which embrace the relation to the world and the world itself, and which make irrelevant the factors that vary from individual to individual."[4] Thus, it becomes possible, in describing conduct, to take as the point of departure not the actor but the environment. For this reason we would emphasize (with Riesman) the necessity of considering our fellow actor as *also,* and indeed *primarily,* motivated *from outside*—a more fruitful approach than one where conduct is considered as flowing from inside—from inner principles, convictions, stable character features, etc.

There is undoubtedly a danger that the pervasive bureaucratization and reification of society will have the same effects as the mechanization of manual labor. At the point where one can go on performing in a specialized and automatized fashion without the intervention of mind, the impulses and capacities of the person remain unsatisfied, and the person himself becomes estranged from those aspects of social existence relating to labor. This is a new aspect, focused on the individual personality, of the massive restructuring of social arrangements, and its specific import is that of converting character into no more than (in the words of Bürger-Prinz), "a place wherein something occurs under the pressure of metaindividual tendencies."

In the realm of industry this has had very considerable consequences: as the advance of mechanization made mind dispensable, the trained, skilled worker was replaced by the semi- or unskilled; and as a side-effect of this the interests of the mind were shifted away from the sphere of work. Reduction in the length of the working day due to technological advance strengthens a trend originating from internal causes: the worker is encouraged even more to seek a content for his own existence outside of work. Previously this development magnified the significance of political concerns; now it finds expression in the formal and empty notion of "leisure." Alfred Weber had already identified this phenomenon nearly fifty years ago: the worker distances himself from his mentally and sensuously impoverished work; he separates the spiritual center of his existence from work, and seeks instead in "life," particularly political life, what work totally or partially denies him: personality and freedom. Weber also saw that this process would inexorably extend its effects to the middle and upper strata; their lives, too, would come to depend upon "dead mechanism." He did not foresee, however, that they would also risk being replaced by semiskilled workers.

Alfred Weber voiced a fear that these latter individuals might not do what the manual workers had done, and instead of distancing themselves from the mechanism (*Apparat*) they would "consider the existence it offers them as *the* life, treat the narrow, specialized, tightly controlled activity it assigns them as *the* activity, and the advance of the mechanism as the central aspect of their own existence."

Weber saw only one response to such a danger: "We must seek to preserve ourselves from the mechanism, to remain human beings, persons, living forces. . . . In evaluating an individual, we should not inquire into his job and what he does at it, but rather how he acquits himself of it, whether he surrenders to it or remains spiritually free of it while performing it, thus remaining spiritually alive."[5]

Brief quotes sometimes formulate the issue too sharply. What Weber, with some reason, feared, was indeed a gigantic and all-pervasive machinery, the frightening image of a pseudoliving machine which grows upon itself, under the control of obsessive pedants or of the contemporary successors to those petty deities the Egyptians or Chinese saw in charge of this or that locality or function. Yet, distancing oneself internally from the machine or (what is the same) from one's occupation, or freeing oneself in spirit from one's work, is only possible insofar as one turns for something spiritually vital to realms of culture which escape the routinization of everyday existence. However, he who puts his leisure to cultural uses often does so only as an exercise in aesthetic appreciation; this generates what Gervinus has called "a fine spiritual egoism," centered around interests which, no matter how sublimated, are but consumer interests of a loftier kind.

One must also take into account another phenomenon. There is a wide stratum of executive or higher managerial personnel in politics, the economy, higher education, administration, the judiciary, as well as in the free professions (lawyers, physicians) which has not yet benefited from a shortening of the workday, and can go on busying itself—with patients or with papers—until late in the evening. This pattern must be taken into consideration when dealing with Alfred Weber's strategy for salvaging the person's vitality once it has distanced itself from routinized occupational cares: that of undertaking cultural pursuits. It would be ridiculous to argue with an author about a sentence he wrote fifty years previously; but we shall use it as a point of reference in addressing a more general concern.

In essence, what higher culture involves is a century-long process of elaboration and transmission of loftier thoughts and more momentous decisions, a process made possible by these having become embodied in solid forms, which can thus be handed on in spite of the limited capacity of pettier souls. In this fashion such products defy not only time but men themselves. Still, they constitute a "low probability" phenomenon, never assured of its further duration. The mere contemplation of these treasures cannot, by itself, guarantee their continued significance. There are only two ways of preserving cultural achievements: by stereotyping them, or by casting them in ever-new forms. In the first case they can preserve their social significance insofar as they attain a monopoly position, as in the cases of various religions, the concepts of Roman law, or certain stylistic devices developed by the Egyptians, which came to be considered as unquestionably valid for all times and situations. In the second case cultural products preserve their validity in the presence of competing tendencies by changing their forms; this, to use contemporary examples, is the problem the theater has not yet solved vis-à-vis film, and the problem that painting, under pressure from photography, solved by surrendering realism.

Equally, there are two ways of destroying or hollowing out cultural forms: as a matter of intentional policy, or by their losing their social role. The latter phenomenon may come about insofar as given cultural forms become exclusively so many stimuli for individual and in particular for aesthetic interests; as they are transformed into delicacies. In painting, for instance, the ancients considered the principle of the imitation of nature obvious, and whoever sought instead to ennoble existent realities had to justify his works as either beautiful or sublime: these were the basic aesthetic categories, and they denoted something visible, which emerges from nature and surmounts it. Those same categories lose credibility with the end of agrarian culture, and thus of the feudal era. In contemporary painting we no longer see at work either the religious devotion of the Middle Ages, or the hunger for glory of the Renaissance princes, or the baroque rejoicing in existence, nor finally the "idealism" of a highly cultivated, apolitical bourgeoisie. Painting, though it is unavoidably moving in the direction of becoming pure decoration, as did Pompeian or late-baroque painting, hesitates to commit itself entirely. The sophisticated artistic themes of objectless painting, the heightened feeling-states of surrealism, can be shared by only few people, and even for those they amount only to a particularly exquisite form of consumer goods. An art previously secure in its secular tradition has become thoroughly "transformed"; it is incomprehensible that art schools should still teach drawing, since such art no longer has anything to offer the amateur.

The specialization to which the "culture carriers" themselves become committed further qualifies the position of culture and cultivation in contemporary society. All specialism perforce tends to generate one-sidedness and habituation, with ambivalent consequences: On the one hand, within the boundaries of a given tendency, performance can become more and more accomplished (only specialists can become virtuosos); on the other, as we have already indicated, habituation makes those factors idiosyncratic to a given personality relatively irrelevant, and the one-sidedness of those remaining relevant aspects makes the properly individual, "indivisible" element in him insignificant. One can thus speak (as Marx did in *The German Ideology*) of a difference which emerges between the existence of the individual, insofar as it is personal to him, and his existence insofar as he is subsumed under this or that branch of labor and under the relative conditions of that labor. The last quotation from Alfred Weber points to that difference. As far as a person's reified and habitualized aspect is concerned, including his standardized emotions, convictions, and thought patterns, the operating habits of each "role holder" cannot differ very much from those of whoever stands next to him at the machine; the extent of that agreement, functional as it is from a certain viewpoint, on the other hand betrays the pressure of the automatisms internal to each, as revealed by the respective tendencies to rigidity, to the avoidance of self-criticism and self-control. Under such conditions the personality becomes absorbed by the different sets of machinery; it is reduced to a *residuum personale,* most obviously when the specialist specializes in imparting expertise. Ortega y Gasset saw this, and expressed it with somewhat excessive sharpness:

Anyone who wishes can observe the stupidity of thought, judgment and action shown today in politics, art, religion, and the general problems of life and the world by the "men of science," and of course behind them, the doctors, engineers, financiers, teachers and so on. . . . They symbolize, and to a great extent constitute, the actual dominion of the masses, and their barbarism is the most immediate cause of European demoralization.[6]

However overstated, his view contains a kernel of truth; one can at any rate legitimately contend that the specialized functionary, endowed with the specialized higher training so quickly acquired today, provides *no* defense against the relapse into barbarism.

9 Personality

MANY VOICES HAVE been heard, of late, expressing concern over the fate of the autonomous individual. Many authors argue that the prospects for the emergence and preservation of personality are steadily worsening—and little wonder, considering that, as Riesman says in one of those asides of his that sound like pistol shots, "the rich as well . . . have inhibited their claims for a decent world."[1] We should here, going on from the argument in chapter 8, like to explore some aspects of the complex question of personality and its preservation in the contemporary situation.

The common view that mass culture threatens the personality is only half correct. Elsewhere, we have argued that there has never been in the world as much differentiated and articulated *subjectivity* as today.[2] Evidence for this is offered for instance by the contemporary arts as a whole; by the public's inexhaustible willingness to lend benevolent attention to the most extravagant displays of subjectivity; or by the phenomenon described above as "the new subjectivism" (chapter 4). Finally, we now live in the era of psychology, or perhaps are barely emerging from it. Over the last century the infinite diversity of individual particularities has attracted and rewarded the interests of psychologists, both in the arts and in science. Nor should one forget that even at the level of the larger society an emphasis on the lived immediacy of the personality has come to constitute a principle for the creation of persistent relations and reciprocities. Fewer and fewer social relations are controlled by obligations determined by one's membership in an estate, corporation, or occupation; rather, the tendency is for people to form groups of a non-official nature outside the public sphere. There become formed, "under the table" as it were, a wide variety of "informal" associations—of restricted, anonymous, and yet often significant networks of friendships, ties of trust, and commonalities of feeling. Such groupings are known only to those involved in them, are not registered outside the reach of one's first-hand experience, and become formed on entirely subjective grounds, on the strength of personal goodwill.

The era of the masses, then, is also an era of narrow special groupings based on personal commitment—a pattern of group formation which would have been unthinkable in the Middle Ages or in the baroque era. Since they are so thoroughly subjective in their constitution, it is hard to determine, for such groupings, how influential they are or how stable. Their peculiar sociological interest also lies in the fact that they do not easily lend themselves to investigation through the standard research device of the survey.

Thus, it is precisely in the larger societies of our mass civilization that the personality-as-subject finds its place as (as it were) the precipitate of those societies. But one must take into account a further meaning of the word "personality," designating the nonroutine individual, or rather the individual who possesses a larger routine and who is capable of transcending it. This "personality" is much in demand in the economy, politics, administration; the "ideal type" of such a figure is that of an individual endowed with vitality and energy, intelligence and capacity for detachment, decisiveness and initiative, inventiveness and discretion—success personified, as it were. Such a type is required and generated by modern society's open-endedness, by its intrinsic tendency to discard traditions and move forward to new targets, and at the same time by its need to counter the threat of rigidity implicit in the sway held by specialized activities. On all these accounts, the individual who raises himself above routines and breaks through them by mastering them, constitutes a key figure. Undoubtedly such a normative type exists in large numbers, in spite of the obstacles opposing its manifestation and its success, and in spite of the punishing fate befalling those who overreach themselves and fail their target.

These two determinations of the notion of "personality" do not exhaust it; one must add personality in the elusive meaning of the significance attaching to what is qualitatively uncommon. Personality in this meaning cannot preserve itself by totally rejecting the machinery of social existence, and operating exclusively as the vessel of "cultural values." Thus understood, personality becomes throughout a matter of discriminating enjoyment (*Genuss*) and while it would be too puritanical to deny that enjoyment is a source of energy, it cannot possibly constitute a program. Besides, the kind of cultivation (*Bildung*) which we still associate with this imagery cannot be preserved once it has become a specialistic pursuit; the "barbarism of reflection" (Vico) is pervasive, and the half-cultivated, trained specialist, as the most recent history shows, is no guarantor of freedom. No matter how indispensable the expert may become in a rationalized society, his own perspective (which originated within the world of crafts and agriculture, and then moved on to prosper in that of industry) is of no ultimate significance. Rational goals hold sway over much of the surface of society; man has become the object of administrative activity in the most intimate aspects of his existence; and a rapidly acquired technical competence is sufficient to give rational expression to this imbalance of power. But it is a much more difficult, much more personal achievement to express coherently a faith in extrarational values, to confront responsibly the tasks set for one by history, to motivate oneself to a more discriminating and discrete performance of rational tasks. Something of this kind, for instance, used to constitute the moral appeal of politics in its higher forms. The tacit assumption of certain inalienable premises operated as a standard for selection from among lines of action originally worked out exclusively in terms of their functional suitability.

In modern society, institutions have become reduced to what is functionally suitable or rather what is held to be suitable; and this, among other reasons, because beyond certain dimensions situations have to be simplified if one is to change them. But the study of past cultures (including that of classical antiquity) convincingly demonstrates the meaningful, the variously symbolic, significance of institutions; in the past these were arrangements with multiple goals, and for that very reason they were oriented to something more than goals. The crucial aspect of a lasting institution is its overdetermination: it must not only prove useful and functional in a proximate, practical sense, but operate also as the point of reference and the "behavior support"* of higher interests; it must allow the most demanding and noblest motivations to express themselves. Under these conditions the institution can fulfill the deeper needs (both material and spiritual) of men, those for conti-

*In English in the text.

nuity, security, and commonality; indeed, it can make possible something like happiness, if by happiness is meant the condition of not being alone in seeking self-transcendence.

Unless they have become exclusively establishments for fueling the individual's effort to attain higher social status, educational institutions occupy a special position, in that they are unavoidably concerned with things supposed to be intrinsically valuable. Because they are also connected with the realm of practical applications, these institutions make viable such vulnerable values as freedom and cultivation, whereas conversely these values, insofar as they are upheld, do not allow the calculation of useful effects to become an end in itself. On this account, to live outside institutions entails for man "the estrangement which consists in possessing consciousness in two separate worlds" (Hegel).

As society disintegrates in the industrial era, the place of institutions is taken by organizations. This entails that there is no longer any constraint upon the goals one may set oneself, and arbitrary action hugely extends its provinces. As soon as a force gains the upper hand in the unceasing interplay and contraposition of tendencies, it makes the most of the opportunity to alter whatever arrangements exist; hence the feeling that things advance zigzagging at an accelerated tempo. Stable institutions measure what peoples are capable of. What is in question is not whether these or those theories are correct, but whether one can configure them in such a way as to interpenetrate men's mutual needs with their ideals—configure them *juridically*, in the sense of Hegel's concept of law: "Law's great mission: the spirit makes itself real. Nature is what it is."[3] Thus understood, law is a thing at the same time more ideal and more useful than a system of directives imparted to a docile and intimidated people. And as to the capacity for idealism of the human heart, the cheap cynicism with which intellectuals poke fun at it today is not the last word on the subject.

Thus, the salvation of culture cannot be sought alongside, but within the machinery of existence. In its subjective meaning, culture belongs to the person who cultivates a capacity for discrimination and detachment in the face of facts; who refuses to subject himself either to the dictatorship of affects in the realm of the heart or the dictatorship of abstractions in that of the intellect; who is constantly sensitive to the diverse significances of given states of fact, to their unexpressed, potential, untested, vulnerable components. A reasoned optimism is a requirement for the possession of culture, as is, above all, an ability to sustain idealism in human concerns—an ability which constitutes the yet undefined antidote to diffidence: you are willing to let others overtake you because you have assumed all along their ability to do so.

Some readers may agree with the above, and yet feel that it points to an anemic, airy-fairy, ineffective idealism. Such an impression should be taken seriously. When the more exacting spirit is confined in isolation by the prevailing conditions, its impotence is more drastically apparent. This is nobody's fault. Within the structure of reality, according to recent ontology, the higher categories are free and autonomous vis-à-vis the lower ones, but the latter are stronger.[4] Within the realm of social institutions this relationship is obscured to the point that the opposite suggests itself: there the forces of the spirit, good taste, justice, appear sometime to make grosser interests work for their own benefit; they hold fast to the garment of egoism, and allow it to drag them along. Expressed as conduct, as the way in which one perceives and pursues his advantage, they surreptitiously preserve their validity, and at any given moment they can shed this covering and openly reassert their claims. This is apparent in the operations of law—exactly insofar as one pursues his own advantage methodically, consistently, plannedly, one needs to attract the trust of others; and it may appear as if the calculation of one's utility contained from the beginning an undisclosed irrational component, which at the end point reveals itself as—justice. It is impossible to assert one's self-interest lastingly without making any concessions to justice; for this reason justice hangs on tightly to self-interest, and the latter can but push the former along.

Thus, disciplining the mutual conduct of men through law means offering such ideals as freedom and justice a chance of realization. Not the certainty of it, since even the observance of law can become mindless habit, or operate as a cover for deceit. All the same, the possibility of those ideals being realized persists as long as the institution does. To some extent institutions embody the existence and the power of the ideal, which they serve by moving it from the slippery terrain of the subjective to the firm ground of rational facts, needs, and interests.

This is why the ancients erected temples to their lawgivers and the founders of their institutions. Every era has reserved special appellations and honors for such extraordinary, improbable deeds and operations. If we use with emphasis the term "personality" to denote the ability to produce wondrous effects, then in our own time we should not seek for such effects in the spheres of culture, literature, or the arts, but rather wherever someone seeks to affirm the exacting claims of the spirit within the machinery itself of existence, rather than "distancing" himself from it. The person energetic and inventive enough to place the massive forces of everyday life at the service of the higher and more vulnerable values, spirited enough to exploit what even the most routine situations have to offer, and to attune himself to all their qualities—such a person is a personality in this peculiar sense. To perform such feats one must be free from both excessive circumspection and excessive self-confidence, must operate almost by instinct under the pressure of ideals, must keep the situation and oneself under control, and convert such control into action. What is productive is that which is improbable, that which must respond differentially to very complex and changing circumstances. Today what is most improbable is the capacity of expressing from oneself, in one's activity, more themes, more inspirations than the situation requires, than others expect, than others express.

Exactly the "exploitation" of everyday situations is the sole thinkable replacement for the sovereign, final act which the machinery of everyday social existence makes impossible today.

A personality is an institution in *one* instance.

Notes

1. Man and Technique

1. Max Scheler, *Die Stellung des Menschen in Kosmos;* Arnold Gehlen, *Der Mensch.*

2. See note 1; and Arnold Gehlen, *Urmensch und Spätkultur.*

3. Werner Sombart, *Der moderne Kapitalismus;* Pául Alsberg, *Das Menschheitsrätsel;* José Ortega y Gasset, *Ideas y creencias.*

4. Georg Kraft, *Der Urmensch als Schöpfer.*

5. Hans Freyer, *Theorie des gegenwärtigen Zeitalters,* p. 27.

6. Henri Bergson, *Creative evolution.*

7. *Ibid.,* p. 169.

8. Gehlen, *Urmensch und Spätkultur,* ch. 22.

9. Max Weber, *The Protestant Ethic and the Spirit of Capitalism.*

10. Norbert Wiener, *The Human Use of Human Beings.*

11. Maurice Pradines, *L'esprit de la religion.*

12. See Gehlen, *Urmensch und Spätkultur,* chs. 43–45.

13. *Ibid.,* chs. 24, 48.

14. Hermann Schmidt, "Die Entwicklung der Technik," pp. 119 ff.

15. Gehlen, *Der Mensch,* introduction and part 2.

16. *Ibid.,* chs. 13 ff.

17. Schmidt, "Die Entwicklung der Technik," p. 121.

18. Wiener, *Human Use of Human Beings,* p. 68.

19. Gehlen, *Der Mensch,* ch. 6.

20. R. Wagner, "Biologische Reglermechanismen," p. 127.

21. Schmidt, "Die Entwicklung der Technik," p. 119.

22. A. Walther, "Probleme im Wechselspiel," p. 139. The issue of *Zeitschrift des Vereins der deutschen Ingenieure* from which Walther's quotation is taken (March 1953) reports the proceedings of the special session of the Association of German Engineers dedicated to "the transformation of man due to technique."

23. Wagner, "Biologische Reglermechanismen," p. 129.

24. *Ibid.,* pp. 124 ff.

25. E. Holst and H. Mittelstaedt, "Das Reafferenzprinzip," p. 37.
26. Schmidt, "Die Entwicklung der Technik," p. 121.

2. Novel Cultural Phenomena

1. See Albert Einstein and Leopold Infeld, *The Evolution of Physics.*
2. Peter Hofstätter, *Einführung in die quantitativen Methoden der Psychologie.*
3. Walter Toman, *Dynamik der Motive.*
4. Jacob L. Moreno, *The Foundations of Sociometry.*
5. See in this connection two standard works in ethnosociology: Claude Lévi-Strauss, *Elementary Structures of Kinship,* and G. P. Murdock, *Social Structure.*
6. The stimulating work of Ulrich Kahrstedt, *Geschichte der griechisch-römischen Altertums,* is of considerable interest in this context.
7. Gottfried Benn, *Probleme der Lyrik.*
8. S. Giedion, *Architektur und Gemeinschaft,* p. 119.
9. See several photographs in S. Giedion, *Architektur und Gemeinschaft.*
10. Hans Freyer, *Theorie des gegenwärtigen Zeitalters,* pp. 15 ff.
11. Giedion, *Architektur und Gemeinschaft,* p. 113.
12. Vilfredo Pareto, *The Mind and Society.*
13. Oskar Morgenstern, *Jahrbuch für Sozialwissenschaft,* vol. 1, 1950.
14. Giedion, *Architektur und Gemeinschaft,* p. 74.
15. Jacob Burckhardt, *The Age of Constantine the Great,* pp. 219–21.
16. Cited in Peter Hofstätter, *Die Psychologie der öffentlichen Meinung,* p. 105.
17. Misia Sert, *Misia.*

3. Social-Psychological Findings

1. See for instance, L. Shaffer, *The Psychology of Adjustment.*
2. David Riesman, *The Lonely Crowd.*
3. Helmut Schelsky, "Im Spiegel des Amerikaners," p. 373.
4. Gustave Thibon, *Retour au réel,* p. 57.
5. Bronislaw Malinowski, *A Scientific Theory of Culture,* p. 82.
6. John Dewey, *Art as Experience,* p. 137.
7. Riesman, *Lonely Crowd.*
8. Gerhard Mackenroth, *Bevölkerungslehre.*
9. Margret Boveri, *Der Verrat im 20. Jahrhundert.*
10. Wilhelm Röpke, *Explication économique du monde moderne,* "Contexture de la division du travail."

11. Thibon, *Retour au réel,* p. 53.

12. Arnold Ruge, "Die Dichter des Chamisso'schen Musenalmanachs für 1839."

13. José Ortega y Gasset, *The Revolt of the Masses,* p. 78.

14. Peter Hofstätter, *Sozialpsychologie,* pp. 109 ff.

15. *Jahrbuch der öffentlichen Meinung 1947–1955* (Allensbach, 1956), pp. 90, 181.

16. Hofstätter, *Sozialpsychologie,* pp. 79 ff.

17. Boveri, *Der Verrat im 20. Jahrhundert.*

18. Joseph Schumpeter, *Capitalism, Socialism, and Democracy,* p. 287.

19. Röpke, *Explication économique,* p. 122.

20. Peter Heller, "Phantasie und Phantomisierung," p. 921.

21. Hans Freyer, *Theorie des gegenwärtigen Zeitalters,* p. 72.

22. Henri Bergson, *Two Sources of Morality and Religion,* ch. 4.

4. The New Subjectivism

1. Arnold Gehlen, "Industriegesellschaft und Staat," *Wort und Wahrheit* (1956), vol. 9.

2. Ernst Howald, *Die Kultur der Antike,* p. 57.

3. Friedrich Nietzsche, *Gesammelte Werke,* vol. 7, p. 380.

4. Arnold Gehlen, *Urmensch und Spätkultur.*

5. The quoted material is taken from Emerson, *Essays* (New York: Dutton, 1906).

6. Gehlen, *Urmensch und Spätkultur,* pp. 80 ff.

7. Maurice Blondel, *Einführung in die Kollektivpsychologie,* p. 174.

8. Wilhelm Hausenstein, *Kairuan,* p. 24.

9. Mme de Staël, *De l'Allemagne,* vol. 2, p. 28.

10. Arnold Geulinx, *Ethik,* ch. 2.

11. Herman Finer, *The Future of Government,* p. 64.

12. Johan Huizinga, *In the Shadow of Tomorrow,* p. 157.

13. Johan Huizinga, *Homo Ludens,* p. 205.

14. *Frankfurter Allgemeine Zeitung,* October 8, 1955. See also a story in *Der Spiegel* (1956), 42:64, to the effect that the mayor of a town had proposed to the town council that in order to arouse more interest in the coming local elections its results should become the object of a publicly sponsored pari-mutuel betting system, with money prizes, and with the bulk of the betting fees going to the needy at Christmas.

15. See Gottfried Benn, *Die Stimme hinter dem Vorhang; Doppelleben.*

16. Hausenstein, *Kairuan*, p. 95.

17. Wilhelm Worringer, *Fragen und Gegenfragen*, p. 146.

5. The Secular Horizon

1. René Descartes, *Discours de la méthode*, pt. 6.

2. Jacques Maritain, *Le Songe de Descartes*, p. 261.

3. Arnold Gehlen, *Urmensch und Spätkultur*.

4. *Ibid.*

5. Fritz Heichelheim, *Wirtschaftsgeschichte des Altertums*, vol. 1, p. 58.

6. *Confluence* (March, 1952) 1:35.

7. Alexis de Tocqueville, *Democracy in America*, vol. 2, p. 899.

8. See for instance Jacob L. Moreno, *The Foundations of Sociometry*; E. Höhn and C. P. Schick, *Das Soziogram*; and Peter Hofstätter, *Gruppendynamik*.

9. Arnold Gehlen, "Mensch trotz Masse," p. 584.

10. Hans Freyer, *Theorie des gegenwärtigen Zeitalters*, p. 15.

11. Margaret Mead, *And Keep Your Powder Dry*, p. 87.

12. Quoted in Jean Bruhat, *Les Journées de février 1848*, p. 174.

13. James Burnham, *The Managerial Revolution*, p. 167.

14. Emil Ermatinger, *Deutsche Kultur im Zeitalter der Aufklärung*, p. 29.

15. Ferdinand Tönnies, *Kritik der öffentlichen Meinung*, p. 77.

16. Johan Huizinga, *In the Shadow of Tomorrow*, p. 183.

17. Herman Finer, *The Future of Government*, p. 16.

18. Georg Gervinus, *Geschichte der deutschen Dichtung*, vol. 5, p. 368.

19. Quoted from Michel Freund, *Georges Sorel*, p. 370.

20. Henri Bergson, *Two Sources of Morality and Religion*, p. 287.

21. Max Scheler, *Schriften zur Soziologie und Weltanschauungslehre*, vol. 2, p. 145.

22. Arnold Toynbee, *A Study of History*, p. 441.

6. Crises in Cultural Development

1. Pitirim A. Sorokin, *Social Philosophy in an Age of Crisis*, ch. 3.

2. This work first appeared in a Russian periodical in 1869, and was first published in book form in 1871.

3. Spengler expressly refers to this statement of Goethe's in his *Decline of the West*, vol. 2, p. 37.

4. *Ibid.*, p. 97.

5. *Ibid.*, vol. 1, p. 367.

6. Erich Franzen, "Die moderne Epik und die deutsche Öffentlichkeit."

7. The reference is to a book by the painter Max Ernst, *Paramythen*.

8. See for instance the first edition of this book, published as *Sozialpsychologische Probleme in der industriellen Gesellschaft* (1949); Pitirim A. Sorokin, *The Crisis of Our Age: The Social and Cultural Outlook* (New York: Dutton, 1957); Hans Freyer, *Theorie des gegenwärtigen Zeitalters;* not to mention the older literature by De Man, Ortega, Jaspers, etc.

9. James McNeill Whistler, *The Gentle Art of Making Enemies.*

10. Spengler, *Decline of the West,* vol. 1, p. 152.

11. Gehlen, *Urmensch und Spätkultur,* p. 294.

12. Alfred Varagnac, *De La Préhistoire au monde moderne,* p. 35.

13. Max Ernst's series of illustrations, *Histoire naturelle,* stands as a symbolic representation of this cultural process: the transformation of the organic into the inorganic, of flesh into stone.

14. Alfred Weber, *Kulturgeschichte als Kultursoziologie,* p. 399.

15. See Gehlen, *Der Mensch,* pp. 339 ff.

16. See Karl Jaspers, *Leonardo, Descartes, Max Weber,* p. 41.

17. F. Keiter, "Die Naturvölker," p. 673.

7. Social Psychology and Psychoanalysis

1. See Ludwig Marcuse, *Sigmund Freud;* Gustav Bally, *Einführung in die Psychoanalyse Sigmund Freuds.*

2. Sigmund Freud, *Introductory Lectures on Psychoanalysis,* p. 356.

3. Freud, *New Introductory Lectures on Psychoanalysis,* p. 126.

4. William McDougall, *Psychoanalysis and Social Psychology,* p. 17.

5. Freud, *Beyond the Pleasure Principle,* p. 22.

6. Otto Rank, *Der Künstler,* p. 34.

7. *Ibid.*, pp. 52–53.

8. Arthur Schopenhauer, *Die Welt als Wille und Vorstellung* (Leipzig: Brockhaus, 1887), vol. 2, ch. 8.

9. José Ortega y Gasset, *The Revolt of the Masses,* p. 115.

10. Freud, *Beyond the Pleasure Principle,* p. 53.

11. Freud, *New Introductory Lectures,* vol. 1, p. 30.

12. Peter Hofstätter, *Sozialpsychologie,* p. 26.

13. *Ibid.*

14. McDougall, *Psychoanalysis and Social Psychology,* p. 36.

15. Freud, *Totem and Taboo*, p. 72.

16. *Ibid.*, p. 73.

17. Freud, *Beyond the Pleasure Principle*, p. 52.

18. See for example, Freud, *New Introductory Lectures*, lecture 35; or *The Future of an Illusion*.

19. Alfred North Whitehead, *Religion in the Making*, p. 86.

20. Sigmund Freud, *Group Psychology and the Analysis of the Ego*, p. 137.

21. Hofstätter, *Sozialpsychologie*, p. 28.

22. *The Dialogues of Plato*, trans. Benjamin Jowett (New York: Random House, 1937), vol. 1, p. 574.

23. McDougall, *Psychoanalysis and Social Psychology*, p. 18.

24. Carl Friedrich v. Weizsäcker, in *Studium Generale*, vol. 1, no. 1.

8. Automatisms

1. Thorstein Veblen, *The Place of Science in Modern Civilization*, p. 348.

2. The expression is in L. Shaffer, *The Psychology of Adjustment*, p. 176.

3. Ludwig Klages, p. 1334.

4. Hans Bürger-Prinz, "Motiv und Motivation," p. 6.

5. Alfred Weber, "Der Beamte."

6. Ortega y Gasset, *The Revolt of the Masses*, p. 97.

9. Personality

1. David Riesman, *The Lonely Crowd* (New Haven: Yale, 1950), paperback, p. 305.

2. Arnold Gehlen, "Mensch trotz Masse," p. 584.

3. Georg W. F. Hegel, *Randbemerkungen zur Rechtsphilosophie*, para. 29.

4. Nicolai Hartmann, *Der Aufbau der realen Welt*, ch. 55.

Bibliography

Alsberg, Paul. *Das Menschheitsrätsel*. Dresden: Sybillen, 1922.

Anders, Günther. *Die Antiquiertheit des Menschen*, 1956.

Baumgarten, Edward. "Mitteilungen und Bemerkungen über den Einfluss Emersons auf Nietzsche," *Jahrbuch f. Amerikastudien* (1956), vol. 1.

Benn, Gottfried. *Die Stimme hinter dem Vorhang*. Wiesbaden: Limes, 1952.

—— *Doppelleben*. Wiesbaden; Limes, 1950.

—— *Probleme der Lyrik*. Wiesbaden: Limes, 1951.

—— *Über mich selbst*. Wiesbaden: Limes, 1956.

Bergson, Henri. *Two Sources of Morality and Religion*. New York: Holt, 1935.

—— *Creative Evolution*. New York: Holt, 1911.

Blondel, Maurice. *Einführung in die Kollektivpsychologie*. Munich: Fischer, 1948.

Boveri, Margret. *Der Verrat im 20. Jahrhundert*. 4 vols. Reinbek: Rowohlt, 1961–.

Bruhat, Jean. *Les Journées de février 1848*. Paris: P.U.F., 1948.

Burckhardt, Jacob. *The Age of Constantine the Great*. New York: Pantheon, 1949.

Bürger-Prinz, Hans. "Motiv und Motivation," *Der Nervearzt* (1947), 18:6.

Burnham, James. *The Managerial Revolution*. New York: John Day, 1941.

Dewey, John. *Art As Experience*. New York: Milton, Balch, 1934.

Einstein, Albert and Leopold Infeld. *The Evolution of Physics*. New York: Simon & Schuster, 1938.

Ermatinger, Emil. *Deutsche Kultur im Zeitalter der Aufklärung*. (Potsdam: Athenaion, 1935.

Ernst, Max. *Histoire naturelle*. Paris: Berggruen, 1926.

Finer, Herman. *The Future of Government*. London: Allen & Unwin, 1945.

Franzen, Erich. "Die moderne Epik und die deutsche Öffentlichkeit," *Merkur* (1955) 92.
Freud, Sigmund. *Beyond the Pleasure Principle*. London: Hogarth, 1955.
—— *Introductory Lectures on Psychoanalysis*. New York: Norton, 1930.
—— *New Introductory Lectures on Psychoanalysis*. New York: Norton, 1933.
—— *Totem and Taboo*. London: Hogarth 1956.
Freund, Michel. *Georges Sorel*. Frankfurt: Klostermann, 1932.
Freyer, Hans. *Theorie des gegenwärtigen Zeitalters*. Stuttgart: Deutsche Verlagsanstalt, 1955.
Gehlen, Arnold. *Der Mensch, seine Natur und seine Stellung in der Welt*. 5th ed. Bonn: Athenäum, 1955.
—— "Mensch trotz Masse," *Wort und Wahrheit* (1952), 8.
—— *Urmensch und Spätkultur*. Bonn: Athenäum, 1956.
Gervinus, Georg G. *Geschichte der deutschen Dichtung*. Leipzig: Engelmann, 1874.
Geulinx, Arnold. *Ethik*. Hamburg: Meiner, 1948.
Giedion, S. *Architektur und Gemeinschaft*. Reinbek: Rowohlt, 1963.
Hartmann, Nicolai. *Der Aufbau der realen Welt*. Berlin: de Gruyter, 1944.
Hausenstein, Wilhelm. *Kairuan*. Munich: Wolff, 1921.
Heichelheim, Fritz Moritz. *Wirtschaftsgeschichte des Altertums*. Tübingen: Mohr, 1938.
Heller, Peter. "Phantasie und Phantomisierung," *Merkur* (1956), 103:921 ff.
Hellpach, Willi. "Sozialpsychologie." In Werner Ziegenfuss, ed., *Handbuch der Soziologie*. Stuttgart: Enke, 1956.
Hofstätter, Peter. *Die Psychologie der öffentlichen Meinung*. Wien: Braunmüller, 1949.
—— *Einführung in die quantitativen Methoden der Psychologie*. Munich: Barth, 1953.
—— *Gruppendynamik*. Reinbek: Rowohlt, 1957.
—— *Sozialpsychologie*. Berlin: de Gruyter, 1956.
Höhn, E., and E. P. Schick. *Das Soziogram*. Stuttgart: Wolf, 1954.
Holst, E., and H. Mittelstaedt. "Der Reafferenzprinzip," *Die Naturwissenschaft* (1950), 37.
Howald, Ernst. *Die Kultur der Antike*. Potsdam: Athenaion, 1936.
Huizinga, Johan. *Homo Ludens*. Boston: Beacon, 1955.
—— *In the Shadow of Tomorrow*. London: Heinemann, 1936.

Jaspers, Karl. "Max Weber." In *Leonardo, Descartes, Max Weber: Three Essays*. London: Routledge, 1964.

Kahrstedt, Ulrich. *Geschichte des griechisch-römischen Altertums*. Munich: Münchener, 1953.

Keiter, F. "Die Naturvölker." In Werner Ziegenfuss, ed., *Handbuch der Soziologie*. Stuttgart: Enke, 1956.

Klages, Ludwig. "Stammbegriffe der Charakterkunde," *Universitas* (Nov. 1947), 1334 ff.

Kraft, Georg. *Der Urmensch als Schöpfer*. Tübingen: Mathiesen, 1948.

Lévi-Strauss, Claude. *Elementary Structures of Kinship*. Boston: Beacon, 1969.

Mackenroth, Gerhard. *Bevölkerungslehre*. Berlin: Springer, 1953.

Malinowski, Bronislaw. *A Scientific Theory of Culture*. Chapel Hill: North Carolina University Press, 1944.

Maritain, Jacques. *Le Songe de Descartes*. Paris: Correa, 1932.

McDougall, William. *Psychoanalysis and Social Psychology*. London: Methuen, 1936.

Mead, Margaret. *And Keep Your Powder Dry*. New York: Morrow, 1942.

Moreno, Jacob L. *The Foundations of Sociometry*. New York: Beacon, 1947.

Murdock, George P. *Social Structure*. New York: Macmillan, 1949.

Ortega y Gasset, José. *The Revolt of the Masses*. New York: Norton, 1932.

—— *Ideas y creencias*. Madrid: Revista de Occidente, 1949.

Pareto, Vilfredo. *The Mind and Society*. New York: Harcourt, Brace, 1942.

Pradines, Maurice, *L'Esprit de la religion*. Paris: Aubier, 1941.

Rank, Otto. *Der Künstler*. Wien: Heller, 1907.

Riesman, David. *The Lonely Crowd*. New Haven: Yale University Press, 1950.

Röpke, Wilhelm. *The Gesellschaftskrisis der Gegenwart*. Zurich: Rentsch, 1948.

Sartre, Jean-Paul. *Situations*. Paris: Gallimard, 1947.

Scheler, Max. *Die Stellung des Menschen im Kosmos*. Darmstadt: Reichl, 1928.

—— *Schriften zur Soziologie und Weltanschauungslehre*. Leipzig: Reinhold, 1923.

Schelsky, Helmut. "Im Spiegel des Amerikaners," *Wort und Wahrheit* (1956), 5:373 ff.

Schmidt, Hermann. "Die Entwicklung der Technik als Phase der Wandlung des Menschen," *Zeitschrift des Vereins der deutschen Ingenieure* (March 1953).

Schopenhauer, Arthur. *Die Welt als Wille und Vorstellung.* vol. 2. Leipzig: Brockhaus, 1887.

Schumpeter, Joseph. *Capitalism, Socialism, and Democracy.* London: Allen & Unwin, 1946.

Sedlmayr, Hans. *Die Revolution der modernen Kunst.* Reinbek: Rowohlt, 1962.

Shaffer, L. *The Psychology of Adjustment.* Boston: Houghton Mifflin, 1936.

Sombart, Werner. *Der moderne Kapitalismus.* Munich: Duncker & Humblot, 1927.

Sorokin, Pitirim. *Social Philosophy in an Age of Crisis.* Boston: Beacon, 1950.

Spengler, Oswald. *The Decline of the West.* New York: Knopf, 1926–1928.

de Staël, Mme A. L. G. *De l'Allemagne.* Weimar: Kiepenheurer, 1913.

Thibon, Gustave. *Retour au réel.* Paris: Laffont, 1944.

Tocqueville, Aléxis de. *Democracy in America.* London: Fontana, 1971.

Toman, Walter. *Dynamik der Motive.* Munich: Fischer, 1954.

Tönnies, Ferdinand. *Kritik der öffentlichen Meinung.* Berlin: Springer, 1922.

Toynbee, Arnold. *A Study of History.* vol. 1. London: Oxford University Press, 1946.

Varagnac, Alfred. *De La Préhistoire au monde moderne.* Paris: Calman-Lévy, 1954.

Veblen, Thorstein. *The Place of Science in Modern Civilization.* New York: Huebsch, 1919.

Wagner, R. "Biologische Reglermechanismen," *Zeitschrift des Vereins der deutschen Ingenieure* (March 1953).

Walther, A. "Probleme im Wechselspiel von Mathematik und Technik," *Zeitschrift des Vereins der deutschen Ingenieure* (March 1953).

Weber, Alfred. "Der Beamte." In *Ideen zur Staats- und Kultursoziologie.* Tübingen: Mohr, 1927.

—— *Kulturgeschichte als Kultursoziologie.* Leiden: Sijthoff, 1935.

Weber, Max. *The Protestant Ethic and the Spirit of Capitalism.* New York: Scribner's, 1930.

Whistler, James McNeill. *The Gentle Art of Making Enemies.* New York: Lovell, 1890.

Whitehead, Alfred North. *Religion in the Making.* New York: Macmillan, 1926.

Wiener, Norbert. *The Human Use of Human Beings.* Boston: Houghton, Mifflin, 1950.

Worringer, Wilhelm. *Fragen und Gegenfragen.* Munich: Piper, 1956.

Index

DATE DUE